T0403091

Milestones in Drug Therapy

For further volumes:
http.//www.springer.com/series/4991

Phil Skolnick
Editor

Glutamate-based Therapies for Psychiatric Disorders

 Springer

Volume Editor
Phil Skolnick, Ph.D., D.Sc. (hon.)
Director, Division of Pharmacotherapies &
Medical Consequences of Drug Abuse
National Institute on Drug Abuse
6001 Executive Boulevard
Bethesda, MD 20892
USA
Phil.Skolnick@nih.gov

Series Editors
Prof. Dr. Michael J. Parnham
Director of Science & Technology
MediMlijeko d.o.o.
Pozarinje 7
HR-10000 Zagreb
Croatia

Prof. Dr. Jaques Bruinvels
Sweelincklaan 75
NL-3723 JC Bilthoven
The Netherlands

ISBN 978-3-0346-0240-2   e-ISBN 978-3-0346-0241-9
DOI 10.1007/978-3-0346-0241-9

Library of Congress Control Number: 2010935790

Cover illustration: NMDA receptor hypofunction model for schizophrenia (see chapter "Activation of group II metabotropic glutamate receptors (mGluR2 and mGluR3) as a novel approach for treatment of schizophrenia" by Douglas J. Sheffler and P. Jeffrey Conn)

Cover design: deblik, Berlin

Printed on acid-free paper

Springer Basel AG is part of Springer Science + Business Media (www.springer.com)

Introduction

Glutamate and Psychiatric Disorders: Evolution Over Five Decades

Studies linking glutamatergic dysfunction to psychiatric disorders preceded the general acceptance of glutamate as a neurotransmitter [1]. Thus, over 50 years ago, Luby et al. [2] reported that administration of the dissociative anesthetic, phencyclidine, to symptom-free schizophrenics resulted in a recrudescence of both positive and negative symptoms. Almost a quarter century elapsed before it was demonstrated that both ketamine and phencyclidine [3] blocked transmission at the NMDA subtype of ionotropic glutamate receptor. These findings led to glutamate-based theories of schizophrenia (reviewed in [4, 5]) and ultimately, the demonstration of the antipsychotic actions of glutamate-based agents [6, 7]. Nonetheless, the use of glutamate-based strategies to treat psychiatric disorders has lagged behind the development of glutamate-based agents in neurological disorders, including stroke and traumatic brain injury [8, 9]. This emphasis was based on compelling evidence that excessive glutamate is neurotoxic and that blockade of ionotropic glutamate receptors attenuates neuron death both *in vitro* and *in vivo* [10, 11]. While not without controversy [12], the evidence that NMDA antagonists reduced ischemic brain damage in models of stroke and traumatic brain injury proved particularly compelling, catalyzing efforts by most major pharmaceutical (and many biotechnology) companies to develop medications to reduce ischemic brain damage. This effort resulted in multiple clinical trials of NMDA antagonists in stroke and traumatic brain injury [8, 9, 13]. However, despite the enormous expense and effort in these indications, there have been no reports of successful trials. This failure may be as much attributable to the difficulties in executing a clinical trial in, for example, stroke, as a failure of the mechanism.

In contrast, preclinical and clinical studies targeting a variety of psychiatric disorders have proceeded in a more deliberate fashion. This pace has in some fashion increased the probability of technical success, due in part to a better understanding of the molecular biology and cellular physiology of glutamate receptors, and in part to an expanded biological toolbox including improved pharmacological agents [14] and the ability to delete or overexpress specific glutamate receptor proteins [15]. Both this enhanced knowledge base and availability of better tools have resulted in remarkable progress on novel approaches to safely modulate glutamatergic function.

Current Approaches to Modulate Glutamatergic Neurotransmission

Glutamate receptors are classified as either "ionotropic" glutamate receptors (iGlu receptors), which are ligand-gated ion channels (NMDA, AMPA, and kainate subtypes) or G-protein-coupled "metabotropic" glutamate (mGlu) receptors (mGlu1–8 subtypes) [16]. Ionotropic glutamate receptors are constituted as heterooligomers, with subunit composition defining both the biophysical (such as relative permeability to cations) and pharmacological (for example, transmitter and ligand affinities) properties of these ligand-gated ion channels. Metabotropic glutamate receptors are composed of homodimers that regulate both second messengers and ion channels, which set the gain of the system to other cellular inputs controlling excitation and plasticity. When metabotropic glutamate receptors are expressed presynaptically, they are capable of regulating glutamate release in either a negative or positive feedback manner. In addition to an array of receptors, the actions of glutamate are regulated by the expression of glutamate transporters (EAAT 1–5), which are expressed in both glial cells and neurons [17]. In glia, mGlu receptors may function to regulate glutamate transporters and calcium signaling involved in glial networks. Ionotropic glutamate receptors also possess multiple modulatory sites that may be exploited to bidirectionally control glutamate neurotransmission [18]. Moreover, the unique role of glycine as an excitatory cotransmitter at NMDA receptors [19] adds yet another potential means of pharmacologically modulating glutamatergic function. Here, in addition to the potential for ortho- and allosteric site modulation, the expression of the glycine transporter GlyT1 tightly regulates how much glycine is available at glutamatergic synapses [20]. In contrast, the glycine transporter GlyT2 is primarily associated with regulating inhibitory glycine neurotransmission at strychnine-sensitive glycine receptors (a ligand gated ion channel), located primarily in spinal cord [21, 22].

Glutamate-Based Therapeutics in Psychiatry

In total, more than 30 proteins that constitute glutamate receptor subtypes and transporters offer a multitude of opportunities for therapeutic intervention. The challenge is to discern the most promising drug targets among these receptors and transporters. In attempting to develop therapeutics for psychiatric disorders, we are presented with behavioral symptoms outside the range of normalcy and pathological brain circuits which may involve developmental insults, genetic and epigenetic influences, and other risk factors. Animal models for psychiatric disorders are problematic because they generally lack face and construct validity. While some animal models possess a high predictive validity (for example, behavioral despair tests for depression) [23], the high failure rate of clinical trials in psychiatric disorders often require multiple trials to appropriately test a hypothesis using a drug acting through a novel mechanism, and thus ultimately lead to validation of these models for novel targets. Nonetheless, over the past decade several promising approaches have emerged based on insights from early stage clinical trials and other pharmacological data in patients challenged with glutamatergic agents.

Thus, early promising positive clinical data have been reported for the mGlu2/3 receptor agonist LY2140023 for psychotic symptoms in patients randomized to placebo, LY2140023 or olanzapine [6]. More recently, Roche announced a positive trial with the GlyT1 transporter inhibitor RO 1678 when combined with atypical antipsychotics in patients exhibiting primarily negative symptoms.

The NMDA antagonist, ketamine, has been shown to have a long lasting antidepressant effect in treatment resistant patients in several clinical trials [24]. These data, consistent with a large body of preclinical evidence, suggest that dampening NMDA receptor function is an effective means of producing an antidepressant action. A recent report [25] documenting that an NR2B antagonist has an antidepressant effect in patients unresponsive to an SSRI provides further support to this hypothesis. In addition, AMPA receptor potentiators, like clinically effective antidepressants, are active in behavior despair animal models and enhance the formation of brain-derived neurotrophic factor [26], which has been linked to the antidepressant actions [27] of biogenic-amine-based agents. More recently, compounds that block mGlu2 receptors have been reported active in these models in an AMPA receptor-dependent fashion. Overall, these promising new approaches are winding their way to clinic for additional validation in humans.

Novel glutamate approaches for anxiety and drug abuse are based on promising data in animal models that have good face validity. In particular, mGlu2/3 receptor agonists were advanced to the clinic based on animal studies and have shown some promise in clinical studies as an anxiolytic [28]. However, the development of these agents has been hampered by potential safety concerns such as seizures. Other approaches such as mGlu2 potentiators and mGlu5 antagonists represent newer strategies that show promise in multiple animals models of anxiety [29]. Likewise, mGlu5 receptor antagonists have been shown to reduce drug seeking behaviors in multiple animal models for opiates, psychomotor stimulants like cocaine and alcohol [30]. Future clinical studies with these agents are planned and should shed light on their clinical promise. This volume describes in detail the bench to bedside approach, developing glutamate-based approaches to treat psychiatry disorders with great unmet need.

Disclaimer. This manuscript was written by PS in a private capacity. The views presented in this Introduction neither represent the views of, nor are they sanctioned by the National Institutes of Health.

Darryle D. Schoepp

Neuroscience Research, Merck and Company, Inc., North Wales, PA

Phil Skolnick

National Institute of Drug Abuse, National Institutes of Health, Bethesda, MD

References

1. Watkins JC, Evans RH (1981) Excitatory amino acid transmitters. Annu Rev Pharmacol Toxicol 21:165–204
2. Luby E, Cohen B, Rosenbaum G, Gottlieb J, Kelley R (1959) Study of a new schizophrenomimetic drug sernyl. AMA Arch Neurol Psychiatry 81: 363–369
3. Anis NA, Berry SC, Burton NR, Lodge D (1983) The dissociative anesthetics, ketmaine and phencyclidine, selective reduce excitation of central mammalian neurons by N-methyl-D-aspartate. Br J Pharmacol 79:565–575
4. Javitt DC, Zukin SR (1991) Recent advances in the phencyclidine model of schizophrenia. Am J Psychiatry 148:1301–1308
5. Conn PJ, Tamminga C, Schoepp DD, Lindsey C (2008) Schizophrenia: moving beyond monoamine antagonists. Mol Interv 8:99–107
6. Patil ST, Zhang L, Martenyi F, Lowe SL, Jackson KA, Andreev BV, Avedisova AS, Bardenstein LM, Gurovich IY, Morozova MA, Mosolov SN, Reznik AM, Smulevich AB, Tochilov VA, Johnson BG, Monn JA, Schoepp DD (2007) Activation of mGlu2/3 receptors as a new approach to treat schizophrenia: a randomized phase 2 clinical trial. Nat Med 13:1102–1107
7. Javitt DC (2004) Glutamate as a therapeutic target in psychiatric disorders. Mol Psychiatry 9:984–997
8. O'Neill MJ, Lees KR (2002) Stroke. In: Lodge D, Danysz W, Parsons CG (eds) Ionotropic glutamate receptors as therapeutic targets, F.P. Graham Publishing Co., Biomedical Book Series Johnson City, TN, pp 403–446
9. Muir KW (2006) Glutamate-based therapeutic approaches: clinical trials with NMDA antagonists. Curr Opin Pharmacol 6:53–60
10. Choi DW, Rothman SM (1990) The role of glutamate neurotoxicity in hypoxic-ischemic neuronal death. Annu Rev Neurosci 13:171–182
11. Gill R, Foster AC, Woodruff GN (1988) Systemic administration of MK-801 protects against ischemic-induced hippocampal neurodegeneration in the gerbil. J Neurosci 7:3343–3349
12. Buchan A, Pulsinelli WA (1989) Are the neuroprotective effects of MK-801 mediated by hypothermia? Stroke 20:148

13. Ikonomidou C, Turski L (2002) Traumatic brain injury. In: Lodge D, Danysz W, Parsons CG (eds) Ionotropic glutamate receptors as therapeutic targets, F.P. Graham Publishing Co., Biomedical Book Series Johnson City, TN, pp 447–466

14. Monn JA, Schoepp DD (2000) Metabotropic glutamate receptor modulators: Recent advances and therapeutic potential. Annu Rep Med Chem 35:1–10

15. Li X, Need AB, Baez M, Witkin JM (2005) Metabotropic glutamate 5 receptor antagonism is associated with antidepressant-like effects in mice. J Pharmacol Exp Ther 319:254–259

16. Pin JP, Acher F (2002) The metabotropic glutamate receptors: structure, activation mechanism and pharmacology. Curr Drug Targets CNS Neurol Disord 1:297–317

17. Maragakis NJ, Rothstein JD (2004). Glutamate transporters: animal models to neurologic disease. Neurobiol Dis 15:461–473

18. Dingledine R, Borges K, Bowie D, Traynelis S (1999) The glutamate receptor ion channels. Pharmacol Rev 51:7–61

19. Kleckner NW, Dingledine R (1988) Requirement for glycine in activation of NMDA-receptors expressed in Xenopus oocytes. Science 241:835–837

20. Aubrey KR, Vandenberg RJ (2001) N[3-(4'fluorophenyl)-3-(4'-phenylphenoxy)propyl]sarcosine (NFPS) is a selective persistent inhibitor of glycine transport. Br J Pharmacol 134:1429–1436

21. Webb TI, Lynch JW (2007) Molecular pharmacology of the glycine receptor chloride channel. Curr Pharm Des 13:2350–2367

22. Liu QR, Lopez-Corcuera B, Mandiyan S, Nelson H, Nelson N (1993) Cloning and expression of a spinal cord and brain-specific glycine transporter with novel structural features. J Biol Chem 268:22802–22808

23. Porsolt RD, Lenegre A (1992) Behavioural models of depression. In: Elliott JM, Heal DJ, Marsden CA (eds) Experimental approaches to anxiety and depression, Wiley, London, pp 73–85

24. Skolnick P, Popik P, Trullas R (2009) Glutamate-based antidepressants: 20 years on. Trends Pharmacol Sci 30:563–569

25. Preskorn SH, Baker B, Kolluri S, Menniti FS, Krams M, Landen JW (2008) An innovative design to establish proof of concept of the antidepressant effects of the NR2B subunit selective N-methyl-D-aspartate antagonist, CP-101,606, in patients with treatment-refractory major depressive disorder. J Clin Psychopharmacol 28:631–637

26. Skolnick P, Legutko B, Li X, Bymaster F (2001) Current perspectives on the development of non-biogenic amine based antidepressants. Pharmacol Res 43:411–423

27. Saarelainen T, Hendolin P, Lucas G, Koponen E, Sairanen M, MacDonald E, Agerman K, Haapsalo A, Nawa H, Aolyz R, Ernfors P, Castren E (2003) Activation of TrkB neurotrophin receptor is induced by antidepressant drugs and is required for antidepressant-induced behavioral effects. J Neurosci 23:349–357

28. Schoepp DD, Wright R, Levine LR, Gaydos B, Potter WZ (2003) Metabotropic glutamate receptor agonists as a novel approach to treat anxiety/stress. Stress 6:189–197

29. Swanson CJ, Johnson MP, Linden AM, Monn JA, Schoepp DD (2005) Metabotropic glutamate receptors as novel targets for anxiety and stress disorders. Nat Rev Drug Discov 4:131–146

30. Linden AM, Schoepp DD (2006) Metabotropic glutamate receptors targets for neuropsychiatric disorders. Drug Discov Today Ther Strateg 3:507–517

Contents

Contributors

P. Jeffrey Conn Department of Pharmacology, Vanderbilt University Medical Center, Light Hall (MRB-IV) Room 1215D, 2215 B Garland Avenue, Nashville, TN 37232, USA, jeff.conn@vanderbilt.edu

Christine M. Gall Anatomy and Neurobiology, University of California at Irvine, Irvine, CA 92697, USA

Hiroshi Kawamoto Tsukuba Research Institute, Banyu Pharmaceutical Co., Ltd, 3 Okubo, Tsukuba, Ibaraki 300-2611, Japan; Medicinal Chemistry, Medicinal Chemistry Laboratories, Pharmaceutical Business, Taisho Pharmaceutical Co., Ltd, 403 Yoshino-cho 1-chome, Kita, Saitama, 331-9530, Japan

Julie C. Lauterborn Anatomy and Neurobiology, University of California at Irvine, Irvine, CA 92697, USA

Gary Lynch Department of Psychiatry and Human Behavior, Gillespie Neuroscience Research Facility, University of California at Irvine, 837 Health Science Road, Irvine, CA 92697-4291, USA; Anatomy and Neurobiology, University of California at Irvine, Irvine, CA 92697, USA, glynch@uci.edu

Athina Markou Department of Psychiatry, School of Medicine, University of California San Diego, 9500 Gilman Drive, M/C 0603, La Jolla, CA 92093-0603, USA, amarkou@ucsd.edu

Eric S. Nisenbaum Neuroscience Discovery, Lilly Research Laboratories, Lilly Corporate Center, Indianapolis, IN 46285-0501, USA, esn@lilly.com

Gabriel Nowak Department of Neurobiology, Institute of Pharmacology, Polish Academy of Sciences, Smętna 12, PL 31-343, Kraków, Poland; Department of Cytobiology, Collegium Medicum, Jagiellonian University, Medyczna 9, PL 30-688, Kraków, Poland; Institute of Pharmacology PAS, Smetna 12, 31-343, Kraków, Poland; Department of Pharmacodynamics, Collegium Medicum, Jagiellonian University, Kraków, Poland, nowak@if-pan.krakow.pl

Hisashi Ohta Research and Development Center, Sato Pharmaceutical Co., Ltd, 6-8-5 Higashi-ohi, Shinagawa, Tokyo, 140-0011, Japan, hisashi.ohta@sato-seiyaku.co.jp

Andrzej Pilc Department of Neurobiology, Institute of Pharmacology, Polish Academy of Sciences, Smętna 12, PL 31-343, Kraków, Poland; Faculty of Health Sciences, Collegium Medicum, Jagiellonian University, Michałowskiego 20, PL 31-126, Kraków, Poland; Institute of Pharmacology PAS, Smetna 12, 31-343, Kraków, Poland; Collegium Medicum, Jagiellonian University, 31-531, Kraków, Poland

Ewa Poleszak Department of Applied Pharmacy, Skubiszewski Medical University of Lublin, Staszica 4, PL 20-081, Lublin, Poland

Piotr Popik Institute of Pharmacology, Polish Academy of Sciences, Smetna 12, 31-343, Kraków, Poland; Faculty of Public Health, Collegium Medicum, Jagiellonian University, Kraków, Poland

Svetlana Semenova Department of Psychiatry, School of Medicine, University of California San Diego, 9500 Gilman Drive, M/C 0603, La Jolla, CA 92093-0603, USA

Douglas J. Sheffler Department of Pharmacology, Vanderbilt University Medical Center, Light Hall (MRB-IV) Room 1215D, 2215 B Garland Avenue, Nashville, TN 37232, USA

Phil Skolnick Division of Pharmacotherapies and Medical Consequences of Drug Abuse, National Institute on Drug Abuse, National Institutes of Health, 6001 Executive Blvd, Bethesda, MD 20892-9551, USA, Phil.Skolnick@nih.gov

Gentaroh Suzuki Tsukuba Research Institute, Banyu Pharmaceutical Co., Ltd, 3 Okubo, Tsukuba, Ibaraki 300-2611, Japan; Central Pharmaceutical Research Institute, Japan Tobacco Inc, 1-1 Murasaki-cho Takatsuki, Osaka, 569-1125, Japan

Bernadeta Szewczyk Department of Neurobiology, Institute of Pharmacology, Polish Academy of Sciences, Smętna 12, PL 31-343, Kraków, Poland

Ramon Trullas Neurobiology Unit, Institut d'Investigacions Biomèdiques de Barcelona, Consejo Superior de Investigaciones Científicas, Institut d'Investigacions Biomèdiques August Pi i Sunyer, Rosselló 161, 08036, Barcelona, Spain Centro de Investigación Biomédica en Red sobre Enfermedades Neurodegenerativas (CIBERNED), Barcelona, Spain

Joanna M. Wieronska Institute of Pharmacology PAS, Smetna 12, 31343, Kraków, Poland

Jeffrey M. Witkin Neuroscience Discovery, Lilly Research Laboratories, Psychiatric Discovery, Lilly Corporate Center, Eli Lilly and Company, Indianapolis, IN, USA, jwitkin@lilly.com

N-Methyl-D-Aspartate (NMDA) Antagonists for the Treatment of Depression

Phil Skolnick, Piotr Popik, and Ramon Trullas

Abstract Depression is a major public health concern that affects $\sim5\%$ of the population in industrialized societies in any given year. Drugs that increase the synaptic availability of biogenic amines (norepinephrine, serotonin, and/or dopamine) have been used to treat depression for over five decades. While the most widely used antidepressants (serotonin and/or norepinephrine selective reuptake inhibitors) are generally safe and effective for many individuals, these drugs are far from ideal. For example, controlled clinical studies have repeatedly demonstrated that ≥2–4 weeks of treatment are required to provide palpable symptom relief. In addition, between 30 and 40% of patients do not respond to a first course of therapy with these biogenic amine-based agents. By contrast, N-methyl-D-aspartate (NMDA) receptor antagonists have been reported to produce rapid and robust antidepressant effects in patients unresponsive to conventional antidepressants. The use of these agents as antidepressants is grounded on a corpus of preclinical evidence, first published 20 years ago, demonstrating the antidepressant-like properties of NMDA antagonists and that chronic treatment with conventional antidepressants attenuates NMDA receptor function. In this chapter, we describe evidence that NMDA antagonists represent an effective alternative to biogenic amine-based agents for treating depression and provide perspective on the hurdles that could impede the development and commercialization of these agents in the face of this remarkable clinical data.

P. Skolnick (✉)
Division of Pharmacotherapies and Medical Consequences of Drug Abuse, National Institute on Drug Abuse, National Institutes of Health, 6001 Executive Blvd, Bethesda, MD 20892-9551, USA
e-mail: Phil.Skolnick@nih.gov

P. Skolnick (ed.), *Glutamate-based Therapies for Psychiatric Disorders*,
Milestones in Drug Therapy, DOI 10.1007/978-3-0346-0241-9_1,
© Springer Basel AG 2010

1 Introduction

Major depressive disorder (depression) is a chronic, recurring illness that affects more than 120 million individuals worldwide. When the impact of depression is measured by the years of healthy life lost to death and disability, by the year 2020 it will be second only to ischemic heart disease as the leading global disease burden [1]. Depression is characterized by the core symptoms of depressed mood and a loss of interest and/or pleasure (termed anhedonia), accompanied by symptoms that may include a significant weight gain or loss, sleep disturbance (either insomnia or hypersomnia), fatigue or loss of energy, a reduction in the ability to think or concentrate, feelings of worthlessness or guilt, recurrent thoughts of death or suicide, and psychomotor agitation or retardation. Based on DSM-IV criteria, a major depressive episode (MDE) is defined by the presence of a core symptom together with four or more other symptoms on a daily, or almost daily, basis for at least 2 weeks [2]. A MDE is invariably accompanied by some degree of social and/or occupational impairment, negatively impacting quality of life and adding to the societal burden associated with lost productivity and increased health care costs.

Converging lines of evidence, including twin and family studies, indicate that genetics significantly contribute to the risk of depression [3, 4], with some estimates suggesting that roughly half of this risk is heritable [5]. Nonetheless, despite extensive candidate gene association studies and genome-wide linkage scans [5, 6], no genes contributing to this risk have been definitively identified. Perhaps, this should not be viewed as surprising absent definitive diagnostic markers, symptoms that can appear diametrical opposites (e.g., weight gain or loss; hyper or hyposomnia) and wax and wane over time. In addition, both environmental and drug-induced epigenetic reprogramming capable of producing enduring changes in gene expression [7–9] would not be detected by conventional genetic analyses (e.g., search for single nucleotide polymorphisms).

Over the past 50 years, drugs increasing the synaptic availability of biogenic amines (norepinephrine, serotonin, and/or dopamine) have been available to treat depression. The majority of antidepressants in current use act via a selective blockade of serotonin and/or norepinephrine uptake. Thus, drugs exemplified by serotonin selective reuptake inhibitors (SSRIs) such as fluoxetine and citalopram, and serotonin/norepinephrine inhibitors (SNRIs) such as duloxetine and venla-faxine, have many fewer serious side effects and are arguably as effective as older biogenic amine-based agents such tricyclics and monoamine oxidase inhibitors and tricyclics [10]. Nonetheless, biogenic amine-based ADs possess significant drawbacks that appear to be inherent to drugs acting via this mechanism [11, 12]. Among these drawbacks is the so-called "therapeutic lag," the 2–4 (or more) weeks of therapy required (in the great majority of carefully controlled trials) to produce a clinically meaningful improvement in depressive symptomatology compared with placebo. At a minimum, the failure to perceive relief within several weeks can negatively impact patient compliance and have far more serious consequences for that subset of depressed patients with suicidal ideation ($\sim15\%$ of depressed individuals commit suicide). In addition, it has been estimated that 30–35% of patients do not respond to a first course of therapy, and of the 60–65% responding to treatment, less than half (that is, only ∼30% of the patient population) either achieve remission or become symptom free [12]. Individuals not responding to the initial AD regimen are administered a different agent, with results that are oftentimes modest and incremental, at best. For example, in an NIH sponsored study [13] of depressed patients who had no symptom remission following an SSRI, Rush et al. reported a remission rate of ∼25% following a second, biogenic amine-based agent, with the remaining cohort apparently treatment resistant.

Clearly, there is a need for agents with a more rapid onset of action and higher efficacy (e.g., a higher proportion of responding or remitting patients) compared with the biogenic amine-based ADs in current use. Grounded on preclinical evidence first published 20 years ago [14], multiple clinical studies [15–18] have now validated NMDA receptors as a target for the development of ADs. Moreover, these studies indicate that an AD action produced via blockade of NMDA receptors may overcome the principal limitations of biogenic amine-based agents. In this chapter, we overview preclinical and clinical evidence indicating NMDA antagonists are AD and discuss the prospects for developing this class of compound to treat depression.

2 Preclinical Studies

2.1 NMDA Antagonists Exhibit AD-Like Actions in Preclinical Tests with High Predictive Validity

Studies to test the hypothesis that NMDA antagonists are antidepressant were prompted by the observations of Shors et al. [19] that exposure to inescapable, but not escapable shock disrupted hippocampal long-term potentiation, a phenomenon dependent upon NMDA receptor activation [20, 21]. The inescapable stress paradigm employed by Shors et al. [19] also produces a behavioral syndrome termed "learned helplessness" [22, 23] that is blocked by biogenic amine-based ADs [24]. Based on these findings, it was hypothesized that the pathways subserved by NMDA receptors were also critical in eliciting the behavioral deficits (i.e., learned helplessness) induced by inescapable stressors, and that interfering with these pathways (by using NMDA antagonists) would, like "classical" antidepressants, mitigate these behavioral deficits.

The initial test of this hypothesis used a murine variant of the forced swim test (FST) [25], a "behavioral despair" paradigm that, like learned helplessness, incorporates an inescapable stressor. The FST has a high predictive validity for clinically effective ADs and was not developed based on preconceived notions of a drug's molecular mechanism of action [26]. The compounds examined in this initial study [14] were the prototypic use-dependent channel blocker, dizocilpine (MK-801), a competitive NMDA antagonist (AP-7), and a glycine partial agonist (ACPC) (Fig. 1). Each of these compounds reduced the time spent immobile, a characteristic of "classical," biogenic amine-based ADs like imipramine. These AD-like effects were observed at doses below those which increase motor activity in the open field, which reduces the possibility of a false positive that can be produced by motor stimulation. Over the past 20 years, more than a dozen reports have described the AD-like actions of NMDA antagonists [27–31] in both the forced swim and tail suspension tests – the behavioral despair measures that are generally considered "gold standards" for screening potential ADs (reviewed in Paul and Skolnick [32]). There is a remarkable structural diversity among active compounds, ranging from classical "drug-like" molecules such as dizocilpine and eliprodil to $Mg++$. Perhaps even more compelling than the structural diversity is that compounds reducing activity through each of the described receptive (orthosteric and allosteric) sites on this family of ligand-gated ion channels have been reported to exhibit AD-like qualities. Thus, competitive NMDA antagonists (e.g., AP-7; CGP 37849), channel blockers (e.g., memantine, ketamine, dizocilpine, $Mg++$), NR2B antagonists (e.g., eliprodil; Ro 25-6981), and glycine site antagonists and partial agonists (e.g., 5, 7-dichlorkynurenic acid; ACPC) (Fig. 1) have all been reported as active.

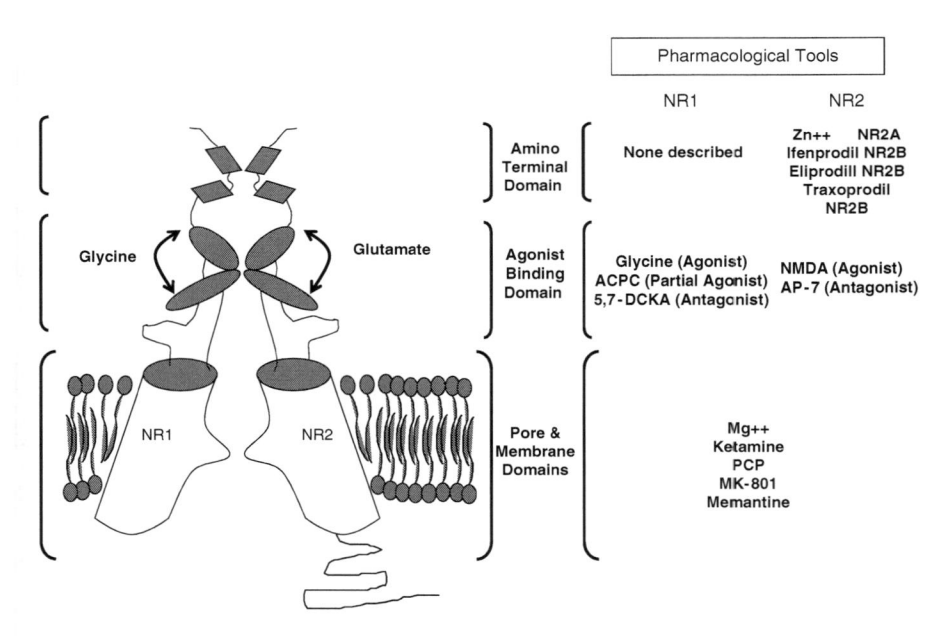

Fig. 1 Schematic representation of an N-methyl-D-aspartate (NMDA) receptor.

NMDA receptors are heteroligomers, with the majority containing NR1 and NR2 subunits. Eight splice variants of the NR-1 subunit and four NR-2 (NR2A-D) subunits have been described, resulting in considerable receptor heterogeneity. Many pharmacological tools, including clinically useful agents (e.g., ketamine, memantine) have been described, allowing for bidirectional modulation of receptor function. The unique requirement for glycine and glutamate as cotransmitters further increases the potential for fine control of receptor function

NMDA antagonists also exhibit AD-like actions in models that possess both greater face and construct validity than behavioral despair procedures routinely employed as drug screens. For example, a variety of NMDA antagonists were reported as active in the chronic mild stress (CMS) model [33–36]. Although many variations of the CMS procedure are currently in use, as originally described by Willner and colleagues (reviewed in Willner [37]), the model is produced by exposing rats to a variety of "mild" inescapable stressors (e.g., cage tilt, stroboscopic lights, wet bedding) that change every few hours. Within a period of weeks, a variety of neurochemical and behavioral changes are observed in rats exposed to CMS [38], including a reduction in the sensitivity to reward, often monitored as a reduction in either the consumption or preference for a palatable solution such as sucrose or saccharin. This reduction is thought to simulate anhedonia (an inability to experience pleasant events), a core symptom of depression. Many of the effects of chronic mild stress, including the reduction in sensitivity to reward, are reversed by chronic (but not acute) AD treatment. Chronic administration of both competitive and noncompetitive NMDA antagonists [33–36] as well as a glycine partial agonist (ACPC) are as effective as imipramine in reversing the deficits in sucrose consumption in this model. Likewise, chronic treatment with NMDA antagonists produce AD-like effects in learned helplessness [39] and olfactory bulbectomy models [40].

2.2   NMDA Receptors Are Altered by Chronic AD Treatment

The marked contrast between the time to achieve steady-state plasma levels (in general, from days to ≤ 1 week for most biogenic amine-based ADs) and the onset of a clinically meaningful effect (in general, ≥ 2–4 or more weeks in most double blind, placebo controlled trials) [41, 42] suggests that neuroplastic change(s) must precede a therapeutic response. Beginning with the pioneering work of Vetulani and Sulser [43], many attempts have been made to describe common neuroadaptive phenomena produced by chronic AD treatments. Certainly, given the range of effective AD treatments, from MAOIs to highly selective serotonin (and/or norepinephrine) transport inhibitors to electroconvulsive shock and sleep deprivation, it could be hypothesized that multiple and perhaps nonconvergent neuroadaptive changes are sufficient to elicit a therapeutic effect. Alternatively, one or more convergent pathways engaged by chronic AD treatments [that is, a common, obligatory pathway(s)] would represent a high value target for novel drug development since compounds acting directly at this target have the potential to effect a more rapid and profound response. During the past two decades, converging lines of evidence indicate that adaptive changes in NMDA receptors are produced by diverse AD treatments. These studies have, in turn, resulted in a stochastic framework [11, 44, 45] resulting in both the development of novel glutamate-based agents [46] as well as stimulating clinical trials examining the AD actions of NMDA antagonists [15].

2.2.1 Neurochemical Studies

A series of studies [47–52] demonstrated that chronic (generally ≥2 weeks), but not single administration of ADs (drugs drawn from every principal class, including ECS) altered the radioligand binding properties of NMDA receptors to rodent brain. Chronic treatment with a more limited series of non-AD drugs (e.g., chlorpromazine, chlordiazepoxide, chlorpheniramine, salbutamol) failed to produce similar changes. The doses selected for this study were based on either reported activity in behavioral despair measures or the ability to downregulate β-adrenoceptors. In these studies, the principal neurochemical measure was the potency of glycine to inhibit the binding of [^{3}H]5,7-dichlorkynurenic acid (DCKA) (a glycine site competitive antagonist) to strychnine-insensitive glycine receptors (also termed glycine$_B$ receptors) that are present on NMDA receptors (Fig. 1). Repeated AD treatment produced between a ~1.8–4.3-fold decrease in the potency of glycine (i.e., an increase in its IC$_{50}$) to inhibit [^{3}H]5,7-DCKA binding to pooled cerebral cortical membranes. A more detailed analysis using a limited number of ADs (including ECS) indicated significant effects were first noted 10–14 days after initiating imipramine treatment and persisted for some time (between 5 and 10 days for imipramine) after cessation of treatment. A dose proportional increase in the IC$_{50}$ of glycine was observed in the two instances where complete dose response studies were performed. Given the role of glycine as a cotransmitter [53] in the operation of NMDA receptor-gated ion channels, AD-induced reductions in the potency of glycine in this neurochemical measure were hypothesized to represent a dampening of NMDA receptor function (reviewed in Skolnick [11]). In addition, other changes in the radioligand binding properties of NMDA receptors were observed following chronic AD treatments. For example, repeated administration of four representative drugs (imipramine, amitryptyline, citalopram, and pargyline) reduced the proportion of high affinity, glycine displaceable [^{3}H]CGP 39653 binding (a competitive NMDA antagonist) to cortical membranes. In the case of citalopram (which produced only a modest increase in the potency of glycine to inhibit [3H]5,7 DCKA binding), a high affinity component of glycine displaceable [^{3}H]CGP 39653 binding was no longer detected [52]. The apparent lack of stoichiometry between these two neurochemical measures together with the presence of other neurochemical changes that were not common to all ADs tested (e.g., reductions in basal [3H]5,7 DCKA, [3H]CGP 39653, and [3H] MK-801 binding) indicates that biogenic amine-based agents are capable of producing multiple effects on NMDA receptors (reviewed in [11]). Given that among ligand-gated ion channels, subunit composition is the primary determinant of ligand affinity and efficacy, a follow-on in situ hybridization study examined the effects of repeated AD treatment on the expression of NMDA receptor subunits. Using citalopram and imipramine as representative agents, Boyer et al. [54] reported that chronic treatment with these drugs reduced the mouse homolog of NR-1 not only in cortex, but in a number of subcortical structures including amygdala and striatum. These reductions were relatively modest (<20%). However, the use of a pan probe in this study may have masked a more robust change in the expression of a particular NR-1 splice variant. Boyer et al. [54] also described a much more robust, albeit neuroanatomically restricted reduction in the mouse homologs of NR-2 subunits. Although both ADs produced a unidirectional (i.e., either a reduction or no change in a particular brain region; in no case was an increase observed) effect on mRNA expression, remarkably, the effects of citalopram and imipramine on NR-2 subunit expression were not identical. This can be exemplified by the very large reduction ($\sim$40%) in the expression of NR2A mRNA in frontal cortex produced by citalopram, with relatively little change in other cortical areas. By contrast, imipramine did not affect expression of NR2A mRNA, but produced significant reductions in NR2B mRNA expression through the cerebral cortex while the effect of citalopram on expression of this subunit mRNA was not statistically significant [11, 54]. Moreover, extensive changes in NMDA subunit mRNA expression were found in subcortical structures [11, 54] that were not detected in radioligand binding studies. However, such differences should not be viewed as surprising given the use of pooled tissues in radioligand binding studies and the greater sensitivity and anatomical resolution of in situ hybridization. Nonetheless, while a stoichiometric relationship between reductions in the expression NMDA receptor subunit mRNA and the AD-induced changes in radioligand binding has not been established, the results obtained with both techniques are consistent with the hypothesis that chronic AD treatment leads to a reduction in NMDA receptor function.

2.2.2 Behavioral and Electrophysiological Studies Confirm That Chronic AD Treatment Blunts NMDA Receptor Function

Behavioral and electrophysiological studies are consistent with the hypothesis that chronic AD treatment blunts NMDA receptor function. Thus, Popik et al. [55] reported that repeated administration of citalopram, imipramine, and ECS reduced the anxiolytic-like effect of L-701, 324 (a glycine site antagonist) in the elevated plus maze (EPM). Moreover, under control conditions, parenteral administration of glycine blocks the effects of L-701,324 in a dose-dependent manner. However, doses of glycine (500–800 mg/kg) that significantly reduced or abolished the effects of L-701,324 in vehicle treated animals were no longer effective in AD-treated mice [55]. The specificity of these effects was confirmed by using a number of positive and negative controls. For example, neither the anxiolytic-like effect of L-701,324 nor the ability of glycine to block this effect was altered in mice receiving: (1) a single dose of imipramine (2) repeated administration of the neuroleptic, chlorpromazine. Further, another important control was the demonstration that chronic treatment with imipramine did not alter the anxiolytic-like action of the benzodiazepine, chlordiazepoxide in the EPM.

Multiple reports have now described a reduction in field potentials from slices of rat frontal cortex following chronic AD administration [56–58]. In these studies, slices were isolated from rat cortex about 2 days after cessation of AD treatment. Imipramine, citalopram, and ECS were reported to reduce amplitude in the more superficial layers (layers II/III) of rat frontal cortex that were evoked by stimulation of layer V. Moreover, a reduction in the amplitude ratio of pharmacologically isolated NMDA to AMPA/kainate receptor mediated components of the field potential were also observed. The authors conclude that chronic treatment with these ADs attenuates glutamatergically mediated synaptic transmission in the cerebral cortex.

The demonstration that conventional ADs produce a time-dependent attenuation of NMDA receptor function was key for the development of a stochastic framework to identify novel AD targets that circumvent the aminergic synapse [11, 44, 46]. Perhaps more important was the realization that if adaptive changes to NMDA receptors preceded a therapeutic response to biogenic amine-based agents, then a direct attenuation of NMDA receptor function might offer a more rapid and/or effective therapy. This hypothesis ([11, 59]; reviewed in [60]) anticipated the remarkable clinical effects of NMDA antagonists described in the following section.

3   Clinical Studies with NMDA Antagonists in Depression

3.1   Multiple Reports Demonstrate a Rapid and Robust AD Response in Patients "Resistant" to Biogenic Amine-Based Agents

Berman et al. [15] first tested the hypothesis that NMDA antagonists are AD by comparing an infusion of ketamine at a subanesthetic dose (0.5 mg/kg) to saline in a small group of medication-free patients. Among the seven patients completing this double blind cross over study, ketamine produced dramatic reductions in the Hamilton Depression Rating Scale (HDRS). These effects were apparent within 3 h and remarkably (given the very short half life of ketamine [$t_{1/2} \sim 2$ h]) could be sustained for at least 3 days [15]. Four of these patients had at least a 50% reduction in the HDRS during the 3-day follow up period in contrast to only one subject infused with saline. Based on both normalization of HDRS scores and clinical impression, the AD effect of ketamine had dissipated within 1–2 weeks. Since ketamine resembles phencyclidine (PCP) in its mode of action at NMDA receptors (Fig. 1), the large, transient (<2 h) spike in both the "high" item of the visual analog scale (VAS) and the positive symptoms component of the brief psychiatric rating scale (BPRS) produced by ketamine was not unexpected. While there was no apparent correlation between reductions in HDRS scores and the manifestation of dissociative effects (as indicated by spikes in the VAS and BRPS), the ability to perceive a drug could be confounding, potentially skewing both patient interviews and investigator ratings. Nonetheless, significant reductions in the suicidality subscale of the HDRS were reported in this small study [15], anticipating a recent report [16] demonstrating that ketamine reduced both implicit and explicit measures of suicidality in a larger study cohort. Zarate et al. [61] reported a similar, rapid (within 2 h) effect of ketamine (0.5 mg/kg) infusion in a larger study (17 patients) of treatment-resistant individuals, defined as patients failing to adequately respond to at least two AD regimens. In this 2006 study, 71% of patients met response (defined as a 50% reduction in HAM-D score) and 29% remission (a HAM-D score ≤ 7) criteria within 24 h of infusion (Fig. 2). One week later, 35% of the patients had maintained a response. This dose of ketamine produced a transient spike in both the BPRS and the Young mania rating scale, which as in the Berman et al. [15] report, could compromise the blinded nature of the study.

In a recent study focused on suicidal ideation in treatment-resistant depression [16], infusion of ketamine (0.5 mg/kg) was reported to produce a highly significant reduction (>2 points; $p < 0.001$) in the suicidality item (with a score of 0 indicating an absence of suicidal ideation to a score of 6 indicating the individual has explicit plans and is actively preparing for suicide) of the MADRAS score (MADRAS-SI) 24 h after infusion, as well as a $\sim$22 point reduction in total MADRAS score ($p < 0.001$). Among the 13 (of 26) patients with clinically significant suicidal ideation (i.e., a score of ≥ 4 on the MADRAS-SI), a remarkable 62% received a rating of 0 or 1 24 h after ketamine infusion, with only two of the patients remaining with scores ≥ 4. In a subset of ten responders, repeated administration of ketamine (an additional five infusions over 14 days) reduced the suicidality index in 9/10 subjects at study end to 0; together with a reduction in total MADRAS score from a mean of 32.7 to 5.1. These data are consistent with the previous reports that ketamine produces a rapid and robust antidepressant action and indicate that this NMDA antagonist may prove useful in acutely suicidal, depressed individuals, for whom conventional AD treatments (including ECS) provide only modest relief over a period of weeks [16].

In a recent open label study, Phelps et al. [17] reported that ketamine (0.5 mg/kg) infusion reduced depressive symptoms in treatment-resistant patients within 4 h.

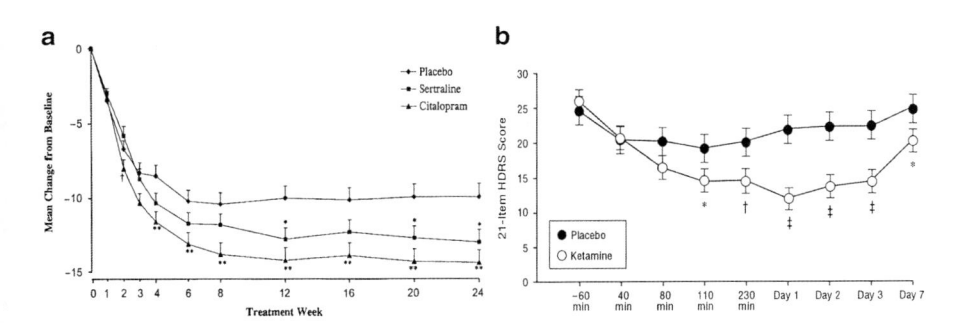

Fig. 2 Comparison of the antidepressant actions of: (**a**) SSRIs and (**b**) ketamine in double blind, placebo controlled settings: Panel **a** represents a double blind, placebo-controlled trial comparing two SSRIs (citalopram and sertraline) in depressed patients. Note that treatment-resistant patients were excluded from this study. Reprinted from Stahl [103], with permission. Baseline HAM-D$_{17}$ scores in this study ranged between 26.4 and 26.6. At endpoint, both SSRIs separated from placebo by 3–4.5 points. Panel **b** is data extracted from a double blind, placebo-controlled study [61] with patients judged resistant to biogenic amine-based agents. Note the difference in time scale between Panels **a** and **b**, and the robust change in HAM-D scale scores manifested within hours after infusion of ketamine

A comparison of patients with a confirmed family history of alcohol abuse to patients with no family history revealed the former group had significantly higher response (67%) and remission rates (42%) than the latter group (18% and 9%, respectively). At a minimum, these findings indicate a family history of alcohol abuse may predict response to the AD actions of ketamine (and perhaps other NMDA antagonists). The NMDA receptor is a target of alcohol action [62], and it could be hypothesized that genetic differences in vulnerability to alcoholism that are linked to NMDA receptors also result in a robust response AD effect of ketamine.

3.2 Studies with Memantine in Depression Are Equivocal

In contrast to the multiple reports of positive ketamine trials in depression, clinical studies with memantine, another use dependent channel blocker with a significantly lower affinity than ketamine [63], have yielded equivocal results. Thus, Zarate et al. [64] reported that oral administration of memantine (5–20 mg/day) failed to separate from placebo ($n = 16$ patients/arm) in an 8-week, double blind, placebo controlled study. These doses of memantine are within the therapeutic range used to treat Alzheimer's disease; in this trial, patients were initially dosed at 5 mg/day increasing by 5 mg/week as tolerated up to the maximum dose. Of note is the use of orally administered medication in this failed memantine study, while all ketamine studies reported to date have used an intravenous infusion. Several reports in the literature have demonstrated that parenterally administered ADs have a rapid onset of action compared with orally administered drug, but other studies have failed to confirm these findings. However, in an open label study [65], memantine produced a significant reduction in depressive symptomatology using the same primary endpoint (MADRAS) as the negative trial. In the successful 12-week study, patients were titrated to 20 mg for 4 weeks, and nonresponsive patients titrated to 30 mg/day at week 8 or 40 mg/day at week 10. While it could be argued that the higher doses of memantine in this protocol might explain the difference between studies, patients improved on both primary and secondary endpoint measures within 1 week of treatment. That is, at the same starting dose used in the failed study. While differences in study design (e.g., open label vs. double blind) could explain the failure of one study and success of another, perhaps of greater importance is the high failure rate of depression trials. For example, among the 39 depression trials filed with the FDA as part of registration submissions in the 1990s, active drug was superior to placebo on all primary and secondary outcome measures less than 15% of the time [66]. There are multiple factors that may contribute to the high apparent failure rate [66], and in view of the positive results obtained with ketamine and traxoprodil (see below), additional clinical trials with memantine merit serious consideration. Nonetheless, a recent report has questioned the widely held view [67] that brain concentrations of memantine achieved after therapeutic doses are sufficient to block the predominant species of NMDA receptors (NR1/2A and NR1/2B) [68].

3.3 A Selective NR2B Antagonist (Traxoprodil) Exhibits AD Activity

Traxoprodil (CP 101,606) is a subtype selective (NR2B) NMDA antagonist that, like many other NMDA antagonists, was initially developed to treat neurological disorders. This compound was entered into clinical trials during the 1990s and failed to improve outcome in a traumatic brain injury study [69, 70]. Preskorn et al. [18] reported a robust AD effect of traxoprodil in patients who did not respond adequately to at least one trial with an SSRI. This study employed an innovative design, incorporating a 6-week open label phase with patients receiving paroxetine and at the midpoint of the open label phase, a single intravenous placebo infusion. Patients who did not respond to paroxetine (defined as less than a 20% reduction in the HAM-D17 scores) during the 6-week period were randomized to groups administered either a single (blinded) infusion of either traxoprodil or placebo, and continued on paroxetine for an additional 4 weeks. Patients receiving traxoprodil had a much larger decrease in the primary endpoint measure (MADRAS scale score at day 5 post infusion) compared with placebo (14.1 vs. 5.5 points, respectively). Furthermore, the response (defined as a 50% reduction in the HAM-D17 scale score) rate to traxoprodil was threefold higher than placebo (60% vs. 20%), and one third of the patients in the traxoprodil score met the criterion for remission (a HAM-D 17 scale score of <7) at day 5 postinfusion. Remarkably, among the 60% of traxoprodil treated patients who met the criterion for response at 5 days, 32% maintained this response at 30 days post infusion. While there is some preclinical literature indicating NR2-B antagonists like traxoprodil would not produce dissociative effects ([71, 72] but see Nicholson et al. [73]), traxoprodil produced moderate to severe dissociative (i.e., phencyclidine like) effects in the first four patients receiving the originally planned dose. While these effects appear to resolve within 6 h, a lower dose was administered to the remaining patients. Nonetheless, this study provides some indication that the manifestation of dissociative symptoms is neither necessary nor sufficient for an AD effect. Thus, among the six patients with dissociative symptoms, two did not meet the response criterion, while among those traxoprodil-treated patients who did not experience dissociative symptoms, more than half met the response criterion. Although this was a small study ($n = 30$), these data indicate an AD response may be achieved without dissociative side effects. Absent dose response data, it is possible that lower doses of traxoprodil may be AD and produce no dissociative effects. This hypothesis will require rigorous clinical testing, with relatively large numbers of subjects. Certainly, the notion that reducing NMDA receptor function through a selective molecular mechanism (in this case, inhibiting NR2B receptors) will provide a better safety profile than nonselective blockade (e.g., with a use dependent channel blocker such as ketamine) has not been proven. It is unlikely that traxoprodil will be developed for depression by Pfizer, since no active trials are currently (January, 2010) listed on www.clinicaltrials.gov.

4   Why Did It Take So Long to Develop NMDA Antagonists for the Treatment of Depression?

Nearly 20 years elapsed between the announcement of an industrial collaboration between Evotec and Roche to develop NMDA antagonists for depression and the demonstration that this class of compound exhibits AD-like actions in well-described preclinical models [14]. By contrast, 13 years elapsed between the first description of fluoxetine as a selective inhibitor of serotonin uptake and FDA approval in 1987 to treat depression. The remarkable and immediate commercial success of fluoxetine followed by other SSRIs throughout the 1990s is perhaps the primary reason that other mechanisms, including NMDA receptor blockade, were viewed as less attractive targets for AD development. Moreover, with converging lines of evidence implicating NMDA receptors as key mediators of excitotoxic cell death [74, 75], many pharmaceutical and biotechnology companies developed NMDA antagonists during the 1990s, most often targeting cerebral ischemia and neurodegenerative disorders [69]. While the potential for NMDA antagonists to produce dissociative effects was known, this was generally viewed as an acceptable risk in an acute, life threatening indication such as stroke. However, absent compelling evidence for a significant clinical advantage over the SSRIs, it would be difficult for scientists in an industrial setting to champion an AD mechanism linked to dissociative side effects. Unfortunately, there have not been any successful trials reported with NMDA antagonists in neurological indications, including stroke and traumatic brain injury [69, 76, 77]. These failures may be as much related to the formidable challenges in the design and execution of clinical trials in, for example, stroke (reviewed in [76]) as the failure of the mechanism to limit ischemic brain damage. Absent a persuasive internal advocate, the failure of a compound (or mechanism) in multiple clinical trials (reviewed in [76]) would tend to diminish corporate enthusiasm for costly trials in other indications. However, with a well characterized molecular target, highly encouraging clinical results, and data indicating that NMDA receptors are a downstream target of biogenic amine-based agents, drug companies may now be set to repurpose (or embark on a synthetic program to develop) NMDA antagonists for depression.

5   Is It Feasible to Develop an NMDA Antagonist for the Treatment of Depression?

While the clinical trials described in the previous section would not meet FDA criteria for registration, the rapid and robust AD effects produced by NMDA antagonists in patients unresponsive to biogenic amine-based agents are compelling. These AD effects of NMDA antagonists are particularly dramatic when compared with a biogenic amine-based agent (Fig. 2). Thus, in a successful placebo controlled trial of two SSRIs, sertraline and citalopram, several weeks of treatment elapse prior to the emergence of a significant separation from placebo, and a three point difference from placebo in HAM-D scale scores is considered clinically significant [78]. By contrast, the response to an infusion of ketamine is manifested within hours, sustained for days, and is significantly more robust (based on HAM-D scale scores) than typically observed with biogenic amine-based agents. Perhaps, even more remarkable is that the subjects in this ketamine study [61] were resistant to treatment with conventional ADs, noting that response rates (generally defined as 50% reduction in HAM-D or MADRAS scale scores) following a 6–8 week trial with biogenic amine-based agents are typically 60–70% compared with 40–50% with placebo [79–81]. These data, together with the need for an alternative to electroconvulsive shock in treatment-resistant depression (TRD), appear to have catalyzed the alliance between Roche and Evotec (announced in March, 2009) to develop NR2B antagonists in TRD.

While the specific target profile of the Evotec/Roche NR2B molecule was not disclosed, formation of this alliance seemingly validates the NMDA receptor as a development target for novel AD. The commercial prospects for a parenterally administered AD are likely limited, particularly if the drug has a potential for producing dissociative side effects. Moreover, the feasibility of developing NMDA antagonists for depression will be determined in large part by issues that have not yet been addressed in the published clinical literature. For example, while both ketamine and traxoprodil appear to possess remarkable AD properties, the efficacy of these agents upon rechallenge is not fully understood. A recent report [16] indicating that reductions in both total MADRAS and suicidality index subscores were maintained after repeated administration of ketamine over a 12-day period is encouraging, because the prospects for future development hinge, in large part, on the maintenance of an AD effect following repeated administration. However, AD efficacy upon rechallenge in a more realistic setting, such as in relapsed patients, will ultimately be a determinant of commercial viability. Also unresolved is the issue of whether repeated administration would lead to sensitization (or desensitization) of potential dissociative side effects, with the former making development problematic and the latter an enabling feature. Clinical resolution of these issues is expensive and will likely be integral to the development plan of any drug candidate with a primary mechanism of NMDA receptor blockade.

In the event that either an orally active NMDA antagonist with pharmaceutically acceptable properties [82, 83] or a parenteral agent lacking psychotomimetic properties can be progressed through clinical testing, there are significant development hurdles that were not in place when ketamine was introduced as an anesthetic in the 1950s. For example, assuming that this agent will be administered in an episodic, subchronic fashion over a lifetime, long-term toxicology (usually 6–12 months in two species) studies will be required. The consequences of long-term NMDA receptor blockade are unknown, but will undoubtedly receive careful scrutiny because of the neuronal vacuolization in retrosplenial and cingulate cortex produced by NMDA antagonists such as MK-801 (dizocilipine) [84]. While neuronal vacuolization is a species-specific phenomenon, these reports nearly halted the development of NMDA antagonists two decades ago for treatment of ischemic insults. Moreover, NMDA antagonists will be evaluated for abuse liability, potential for carcinogenicity, effects on reproduction, and other safety parameters that are requirements for registration.

Perhaps the most problematic issue for the development of an NMDA antagonist is the potential for producing dissociative effects. However, additional studies with traxoprodil and ketamine could yield significant AD effects at doses well below those producing dissociative effects. If the risk of producing dissociative effects is minimized, it may be possible to broaden the use of NMDA antagonists beyond treatment-resistant depression, particularly if an orally active agent can be developed.

An additional strategy for reducing the dose of NMDA antagonist for use as a first or second line therapy is based on an earlier preclinical study [85] combining an NMDA antagonist with an amine-based antidepressant. Thus, Rogoz et al. [85] reported a synergistic effect in the rat forced swim test by combining NMDA antagonists with amine-based agents, including imipramine and fluoxetine. If this synergism also obtains in the clinic, it may be possible to reduce the dose of each agent to yield an effective AD response, reducing or eliminating the most problematic side effects of each agent. This hypothesis could certainly be tested in the clinic, with the added benefit of using two already marketed agents (e.g., memantine and fluoxetine; ketamine and bupropion), circumventing many of the regulatory issues associated with developing a new chemical entity. Nonetheless, given the current very conservative regulatory environment, a high safety bar will be demanded of NMDA antagonists (and other potential antidepressants modulating glutamatergic transmission) as long as safe, albeit less effective, alternatives are available.

6   Why Are NMDA Antagonists AD?: Developing Drugs That Circumvent the Monoaminergic Synapse

Stressful life events can either precipitate or exacerbate mood disorders, including depression (reviewed in Gold and Chrousos [86]). Stress produces a well-described neuronal damage and atrophy that is mediated, at least in part, by glucocorticoids [87–90]. A key observation to understanding the intraclinical pathways engaged by elevating synaptic concentrations of biogenic amines was the demonstration (e.g., [91]; reviewed in Duman et al. [92]) that chronic AD treatments increased the expression of mRNA encoding brain-derived neurotrophic factor (BDNF) in rat hippocampus, which in turn appears to be mediated via an increase in the expression of the transcription factor, cyclic AMP response-element binding (CREB) protein (reviewed in Duman, Nibuya and Vaidya [93]). Acting through its receptor, tropomyosin-related kinase B (TrkB), BDNF exerts both neuroprotective and neurotrophic actions [90, 94–96]. These observations led to the hypothesis that AD-induced increases in BDNF is a pivotal step in blunting the ability of chronic stressors to damage vulnerable neurons [93, 97, 98]. The ability of NMDA antagonists to protect neurons from a wide variety of insults (reviewed in [99]) led to the hypothesis that biogenic amine-based agents and NMDA antagonists converge on a common cellular endpoint, protecting vulnerable neurons against stress-induced damage [11]. However, CREB is capable of activating a large number of genes throughout the CNS, and there is evidence that some of the downstream events consequent to CREB may be manifest in effects that are "prodepressive" [100]. Such diametrically opposed effects produced by increasing the expression (or activation) of a transcription factor like CREB may explain both the delay in onset of biogenic amine-based ADs and a relatively modest therapeutic response in certain individuals.

There is also evidence that BDNF can reduce the expression of mRNA encoding both the NR2A and 2B subunits [101] indicating that both biogenic amine-based agents and NMDA antagonists converge on a common molecular target, leading to a dampening of NMDA receptor function. A corollary of this hypothesis is that agents (NMDA antagonists) acting directly on this target would produce a more rapid action than drugs engaging a more distal target [11]. This corollary has been borne out in clinical trials, and the hypothesis has evolved over the past decade based on evidence that: (1) chronic AD treatments increase synaptic AMPA/NMDA receptor throughput (reviewed in Sanacora et al.[45]); (2) AMPA receptor potentiators, which produce a rapid and robust increase BDNF, are themselves AD [44, 46] (see Chapter by Nisenbaum and Witkin); and (3) the AD properties of ketamine and other NMDA antagonists may be mediated via AMPA receptor activation [29]. This latter finding may have important implications for developing rapid and effective glutamate-based ADs, because if the AD actions of ketamine are the result of AMPA receptor activation, then it may be possible to circumvent the limiting side effects associated with NMDA receptor blockade through agents directly targeting AMPA receptors. However, a subsequent report [102] using a competitive NMDA antagonist indicates that AMPA receptor activation may not be a final common pathway mediating the AD effects of all NMDA antagonists. At face value, these reports appear contradictory and beg a more comprehensive study comparing representative compounds from each class of NMDA antagonist (see Fig. 1). Since both studies appear well-controlled, there may be differences in the mechanisms (e.g., compounds effecting a substantial activation of AMPA receptors in addition to NMDA receptor blockade vs. a compound acting solely through NMDA receptor blockade) NMDA antagonists engage to produce an AD-like response. Probing these differences may help understand how NMDA antagonists effect a rapid and sustained AD action in the clinic.

Note: This manuscript was written by PS in a private capacity. The views presented in this chapter neither represent the views of, nor are they sanctioned by, the National Institutes of Health.

References

1. Murray CJ, Lopez AD (1996) Evidence-based health policy–lessons from the Global Burden of Disease Study. Science 274:740–743
2. American Psychiatric Association (2000) Diagnostic and statistical manual of mental disorders, 4th edn. American Psychiatric Association Press, Washington DC
3. Kendler KS, Eaves LJ, Walters EE, Neale MC, Heath AC, Kessler RC (1996) The identification and validation of distinct depressive syndromes in a population-based sample of female twins. Arch Gen Psychiatry 53:391–399
4. Kendler KS, Davis CG, Kessler RC (1997) The familial aggregation of common psychiatric and substance use disorders in the National Comorbidity Survey: a family history study. Br J Psychiatry 170:541–548
5. Berton O, Nestler EJ (2006) New approaches to antidepressant drug discovery: beyond monoamines. Nat Rev Neurosci 7:137–151
6. Munafo MR, Durrant C, Lewis G, Flint J (2009) Gene X environment interactions at the serotonin transporter locus. Biol Psychiatry 65:211–219
7. Tsankova NM, Berton O, Renthal W, Kumar A, Neve RL, Nestler EJ (2006) Sustained hippocampal chromatin regulation in a mouse model of depression and antidepressant action. Nat Neurosci 9:519–525
8. McGowan PO, Sasaki A, D'Alessio AC, Dymov S, Labonte B, Szyf M, Turecki G, Meaney MJ (2009) Epigenetic regulation of the glucocorticoid receptor in human brain associates with childhood abuse. Nat Neurosci 12:342–348
9. Szyf M, Weaver I, Meaney M (2007) Maternal care, the epigenome and phenotypic differences in behavior. Reprod Toxicol 24:9–19
10. Smith D, Dempster C, Glanville J, Freemantle N, Anderson I (2002) Efficacy and tolerability of venlafaxine compared with selective serotonin reuptake inhibitors and other antidepressants: a meta-analysis. Br J Psychiatry 180:396–404
11. Skolnick P (1999) Antidepressants for the new millenium. Eur J Pharmacol 375:31–40
12. Rosenzweig-Lipson S, Beyer CE, Hughes ZA, Khawaja X, Rajarao SJ, Malberg JE, Rahman Z, Ring RH, Schechter LE (2007) Differentiating antidepressants of the future: efficacy and safety. Pharmacol Ther 113:134–153
13. Rush AJ, Trivedi MH, Wisniewski SR, Stewart JW, Nierenberg AA, Thase ME, Ritz L, Biggs MM, Warden D, Luther JF et al (2006) Bupropion-SR, sertraline, or venlafaxine-XR after failure of SSRIs for depression. N Engl J Med 354:1231–1242
14. Trullas R, Skolnick P (1990) Functional antagonists at the NMDA receptor complex exhibit antidepressant actions. Eur J Pharmacol 185:1–10
15. Berman RM, Cappiello A, Anand A, Oren DA, Heninger GR, Charney DS, Krystal JH (2000) Antidepressant effects of ketamine in depressed patients. Biol Psychiatry 47:351–354
16. Price RB, Nock MK, Charney DS, Mathew SJ (2009) Effects of intravenous ketamine on explicit and implicit measures of suicidality in treatment-resistant depression. Biol Psychiatry 66:522–526
17. Phelps LE, Brutsche N, Moral JR, Luckenbaugh DA, Manji HK, Zarate CA Jr (2009) Family History of alcohol dependence and initial antidepressant response to an N-methyl-D-aspartate antagonist. Biol Psychiatry 65:181–184
18. Preskorn SH, Baker B, Kolluri S, Menniti FS, Krams M, Landen JW (2008) An innovative design to establish proof of concept of the antidepressant effects of the NR2B subunit selective N-methyl-D-aspartate antagonist, CP-101, 606, in patients with treatment-refractory major depressive disorder. J Clin Psychopharmacol 28:631–637
19. Shors TJ, Seib TB, Levine S, Thompson RF (1989) Inescapable versus escapable shock modulates long-term potentiation in the rat hippocampus. Science 244:224–226
20. Harris EW, Ganong AH, Cotman CW (1984) Long-term potentiation in the hippocampus involves activation of N-methyl-D-aspartate receptors. Brain Res 323:132–137

21. Morris RGM, Anderson E, Lynch G, Baudry M (1986) Selective impairment of learning and blockade of long-term potentiation by *N*-methyl-ᴅ-aspartate receptor antagonist, AP5. Nature 319:774–776
22. Seligman ME (1978) Learned helplessness as a model of depression. Comment and integration. J Abnorm Psychol 87:165–179
23. Maier SF, Watkins LR (2005) Stressor controllability and learned helplessness: the roles of the dorsal raphe nucleus, serotonin, and corticotropin-releasing factor. Neurosci Biobehav Rev 29:829–841
24. Leshner AI, Remler H, Biegon A, Samuel D (1979) Desmethylimipramine (DMI) counteracts learned helplessness in rats. Psychopharmacology 66:207–208
25. Porsolt RD, Bertin A, Jalfre M (1977) Behavioral despair in mice: a primary screening test for antidepressants. Arch Int Pharmacodyn Thér 229:327–336
26. Porsolt RD, Lenegre A (1992) Behavioral models of depression. In: Elliott JM, Heal DJ, Marsden CA (eds) Experimental approaches to anxiety and depression. Wiley, London, pp 73–85
27. Kos T, Legutko B, Danysz W, Samoriski G, Popik P (2006) Enhancement of antidepressant-like effects but not BDNF mRNA expression by the novel NMDA receptor antagonist neramexane in mice. J Pharmacol Exp Ther 318:1128–1136
28. Garcia LS, Comim CM, Valvassori SS, Reus GZ, Barbosa LM, Andreazza AC, Stertz L, Fries GR, Gavioli EC, Kapczinski F et al (2008) Acute administration of ketamine induces antidepressant-like effects in the forced swimming test and increases BDNF levels in the rat hippocampus. Prog Neuropsychopharmacol Biol Psychiatry 32:140–144
29. Maeng S, Zarate CA Jr, Du J, Schloesser RJ, McCammon J, Chen G, Manji HK (2008) Cellular mechanisms underlying the antidepressant effects of ketamine: role of alpha-amino-3-hydroxy-5-methylisoxazole-4-propionic acid receptors. Biol Psychiatry 63:349–352
30. Nowak G, Szewczyk B, Pilc A (2005) Zinc and depression. An update. Pharmacol Rep 57:713–718
31. Popik P, Kos T, Sowa-Kucma M, Nowak G (2008) Lack of persistent effects of ketamine in rodent models of depression. Psychopharmacology 198:421–430
32. Paul IA, Skolnick P (2003) Glutamate and depression: clinical and preclinical studies. Ann NY Acad Sci 1003:250–272
33. Papp M, Moryl E (1993) Similar effect of chronic treatment with imipramine and the NMDA antagonists CGP 37849 and MK-801 in a chronic mild stress model of depression in rats. Eur Neuropsychopharmacol 3:348–349
34. Papp M, Moryl E (1993) New evidence for the antidepressant activity of MK-801, a noncompetitive antagonist of NMDA receptors. Pol J Pharmacol 45:549–553
35. Papp M, Moryl E (1994) Antidepressant activity of non-competitive NMDA antagonists in a chronic mild stress model of depression. Eur J Pharmacol 263:1–7
36. Papp M, Moryl E (1996) Antidepressant-like effects of 1-aminocyclopropanecarboxylic acid and D-cycloserine in an animal model of depression. Eur J Pharmacol 316:145–151
37. Willner P (1997) Validity, reliability and utility of the chronic mild stress model of depression: a 10-year review and evaluation. Psychopharmacology (Berl) 134:319–329
38. Willner P, Papp M (1997) Animal models to detect antidepressants. Are new strategies necessary to detect new agents? In: Skolnick P (ed) Antidepressants new pharmacological strategies. Humana Press, Totowa, New Jersey, pp 213–230
39. Meloni D, Gambarana C, De Montis MG, Dal Pra P, Taddei I, Tagliamonte A (1993) Dizocilpine antagonizes the effect of chronic imipramine on learned helplessness in rats. Pharmacol Biochem Behav 46(2):423–426
40. Kelly JP, Wrynn AS, Leonard BE (1997) The olfactory bulbectomized rat as a model of depression: an update. Pharmacol Ther 74:299–316
41. Oswald J, Brezinowa V, Dunleavy DLF (1972) On the slowness of action of tricyclic antidepressant drugs. Br J Psychiatry 120:673–677

42. Manji HK, Drevets WC, Charney DS (2001) The cellular neurobiology of depression. Nat Med 7:541–547
43. Vetulani J, Sulser F (1975) Action of various antidepressant treatments reduces reactivity of noradrenergic cyclic AMP-generating system in limbic forebrain. Nature 257:495–496
44. Skolnick P, Legutko B, Li X, Bymaster FP (2001) Current perspectives on the development of non-biogenic amine-based antidepressants. Pharmacol Res 43:411–423
45. Sanacora G, Zarate CA, Krystal JH, Manji HK (2008) Targeting the glutamatergic system to develop novel, improved therapeutics for mood disorders. Nat Rev Drug Discov 7:426–437
46. Alt A, Nisenbaum ES, Bleakman D, Witkin JM (2006) A role for AMPA receptors in mood disorders. Biochem Pharmacol 71:1273–1288
47. Skolnick P, Layer RT, Popik P, Nowak G, Paul IA, Trullas R (1996) Adaptation of the N-methyl-D-aspartate (NMDA) receptors following antidepressant treatment: implications for the pharmacotherapy of depression. Pharmacopsychiatry 29:23–26
48. Paul IA, Layer RT, Skolnick P, Nowak G (1993) Adaptation of the NMDA receptor in rat cortex following chronic electroconvulsive shock or imipramine. Eur J Pharmacol 247:305–311
49. Paul IA, Nowak G, Layer RT, Popik P, Skolnick P (1994) Adaptation of the N-methyl-D-aspartate receptor complex following chronic antidepressant treatments. J Pharmacol Exp Ther 269:95–102
50. Nowak G, Trullas R, Layer R, Skolnick P, Paul IA (1993) Adaptive changes in the N-methyl-D-aspartate receptor complex after chronic treatment with imipramine and 1-aminocyclopropanecarboxylic acid. J Pharmacol Exp Ther 265:1380–1386
51. Nowak G, Legutko B, Skolnick P, Popik P (1998) Adaptation of cortical NMDA receptors by chronic treatment with specific serotonin reuptake inhibitors. Eur J Pharmacol 342:367–370
52. Nowak G, Li Y, Paul IA (1996) Adaptation of cortical but not hippocampal NMDA receptors after chronic citalopram treatment. Eur J Pharmacol 295:75–85
53. Kleckner NW, Dingledine R (1988) Requirement for glycine in activation of NMDA receptors expressed in Xenopus oocytes. Science 241:835–837
54. Boyer PA, Skolnick P, Fossom LH (1998) Chronic administration of imipramine and citalopram alters the expression of NMDA receptor subunit mRNAs in mouse brain – a quantitative in situ hybridization study. J Mol Neurosci 10:219–233
55. Popik P, Wrobel M, Nowak G (2000) Chronic treatment with antidepressants affects glycine/NMDA receptor function: behavioral evidence. Neuropharmacology 39:2278–2287
56. Bobula B, Tokarski K, Hess G (2003) Repeated administration of antidepressants decreases field potentials in rat frontal cortex. Neuroscience 120:765–769
57. Tokarski K, Bobula B, Wabno J, Hess G (2008) Repeated administration of imipramine attenuates glutamatergic transmission in rat frontal cortex. Neuroscience 153:789–795
58. Bobula B, Hess G (2008) Antidepressant treatments-induced modifications of glutamatergic transmission in rat frontal cortex. Pharmacol Rep 60:865–871
59. Trullas R (1997) Functional NMDA antagonists: a new class of antidepressant agents. In: Skolnick P (ed) Antidepressants new pharmacological strategies. Humana Press, Totowa, New Jersey, pp 103–124
60. Skolnick P, Popik P, Trullas R (2009) Glutamate-based antidepressants: 20 years on. Trends Pharmacol Sci 30:563–569
61. Zarate CA Jr, Singh JB, Carlson PJ, Brutsche NE, Ameli R, Luckenbaugh DA, Charney DS, Manji HK (2006) A randomized trial of an N-methyl-D-aspartate antagonist in treatment-resistant major depression. Arch Gen Psychiatry 63:856–864
62. Lovinger DM, White G, Weight FF (1989) Ethanol inhibits NMDA-activated ion current in hippocampal neurons. Science 243:1721–1724
63. Rammes G, Danysz W, Parsons CG (2008) Pharmacodynamics of memantine: an update. Curr Neuropharmacol 6:55–78

64. Zarate CA Jr, Singh JB, Quiroz JA, De Jesus G, Denicoff KK, Luckenbaugh DA, Manji HK, Charney DS (2006) A double-blind, placebo-controlled study of memantine in the treatment of major depression. Am J Psychiatry 163:153–155
65. Ferguson JM, Shingleton RN (2007) An open-label, flexible-dose study of memantine in major depressive disorder. Clin Neuropharmacol 30:136–144
66. Fava M, Evins AE, Dorer DJ, Schoenfeld DA (2003) The problem of the placebo response in clinical trials for psychiatric disorders: culprits, possible remedies, and a novel study design approach. Psychother Psychosom 72:115–127
67. Parsons CG, Danysz W, Quack G (1999) Memantine is a clinically well tolerated *N*-methyl-D-aspartate (NMDA) receptor antagonist – a review of preclinical data. Neuropharmacology 38:735–767
68. Kotermanski SE, Johnson JW (2009) Mg2+ imparts NMDA receptor subtype selectivity to the Alzheimer's drug memantine. J Neurosci 29:2774–2779
69. Kemp JA, McKernan RM (2002) NMDA receptor pathways as drug targets. Nat Neurosci 5:1039–1042, Suppl
70. Ikonomidou C, Turski L (2002) Traumatic brain injury. In: Lodge D, Danysz W, Parsons CG (eds) Ionotropic glutamate receptors as therapeutic targets. F.P. Graham Publishing Co., Johnson City, TN, pp 447–466, Biomedical Book Series
71. Loftis JM, Janowsky A (2003) The *N*-methyl-D-aspartate receptor subunit NR2B: localization, functional properties, regulation, and clinical implications. Pharmacol Ther 97:55–85
72. Gogas KR (2006) Glutamate-based therapeutic approaches: NR2B receptor antagonists. Curr Opin Pharmacol 6:68–74
73. Nicholson KL, Mansbach RS, Menniti FS, Balster RL (2007) The phencyclidine-like discriminative stimulus effects and reinforcing properties of the NR2B-selective *N*-methyl-D-aspartate antagonist CP-101 606 in rats and rhesus monkeys. Behav Pharmacol 18:731–743
74. Choi DW, Rothman SM (1990) The role of glutamate neurotoxicity in hypoxic-ischemic neuronal death. Annu Rev Neurosci 13:171–182
75. Albers GW, Goldberg MP, Choi DW (1992) Do NMDA antagonists prevent neuronal injury? Yes. Arch Neurol 49:418–420
76. O'Neil M, Lees KR (2002) Stroke. In: Lodge D, Danysz W, Parsons CG (eds) Ionotropic glutamate receptors as therapeutic targets. F.P. Graham Publishing Co, Johnson City, TN, pp 403–446, Biomedical Book Series
77. Muir KW (2006) Glutamate-based therapeutic approaches: clinical trials with NMDA antagonists. Curr Opin Pharmacol 6:53–60
78. Kirsch I, Deacon BJ, Huedo-Medina TB, Scoboria A, Moore TJ, Johnson BT (2008) Initial severity and antidepressant benefits: a meta-analysis of data submitted to the Food and Drug Administration. PLoS Med 5:e45
79. Leber P (2000) The use of placebo control groups in the assessment of psychiatric drugs: an historical context. Biol Psychiatry 47:699–706
80. Quitkin FM, Rabkin JG, Gerald J, Davis JM, Klein DF (2000) Validity of clinical trials of antidepressants. Am J Psychiatry 157:327–337
81. Khan A, Khan SR, Walens G, Kolts R, Giller EL (2003) Frequency of positive studies among fixed and flexible dose antidepressant clinical trials: an analysis of the food and drug administration summary basis of approval reports. Neuropsychopharmacology 28:552–557
82. Suetake-Koga S, Shimazaki T, Takamori K, Chaki S, Kanuma K, Sekiguchi Y, Suzuki T, Kikuchi T, Matsui Y, Honda T (2006) In vitro and antinociceptive profile of HON0001, an orally active NMDA receptor NR2B subunit antagonist. Pharmacol Biochem Behav 84:134–141
83. Liverton NJ, Bednar RA, Bednar B, Butcher JW, Claiborne CF, Claremon DA, Cunningham M, DiLella AG, Gaul SL, Libby BE et al (2007) Identification and characterization of 4-methylbenzyl 4-[(pyrimidin-2-ylamino)methyl]piperidine-1-carboxylate, an orally bioavailable,

brain penetrant NR2B selective N-methyl-D-aspartate receptor antagonist. J Med Chem 50:807–819

84. Olney JW, Labruyere J, Price MT (1989) Pathological changes induced in cerebrocortical neurons by phencyclidine and related drugs. Science 244:1360–1362

85. Rogoz Z, Skuza G, Maj J, Danysz W (2002) Synergistic effect of uncompetitive NMDA receptor antagonists and antidepressant drugs in the forced swimming test in rats. Neuropharmacology 42:1024–1030

86. Gold PW, Chrousos GP (2002) Organization of the stress system and its dysregulation in melancholic and atypical depression: high vs low CRH/NE states. Mol Psychiatry 7:254–275

87. Sapolsky RM (1996) Why stress is bad for your brain. Science 273:749–750

88. Sapolsky RM (2001) Depression, antidepressants, and the shrinking hippocampus. Proc Natl Acad Sci USA 98:12320–12322

89. McEwen BS (2000) Effects of adverse experiences for brain structure and function. Biol Psychiatry 48:721–731

90. Zarate CA Jr, Singh J, Manji HK (2006) Cellular plasticity cascades: targets for the development of novel therapeutics for bipolar disorder. Biol Psychiatry 59:1006–1020

91. Nibuya M, Morinobu S, Duman RS (1995) Regulation of BDNF and trkB mRNA in rat brain by chronic electroconvulsive seizure and antidepressant drug treatments. J Neurosci 15:7539–7547

92. Duman RS, Monteggia LM (2006) A neurotrophic model for stress-related mood disorders. Biol Psychiatry 59:1116–1127

93. Duman RS, Nibuya M, Vaidya VA (1997) A role for CREB in antidepressant action. In: Skolnick P (ed) Antidepressants new pharmacological strategies. Humana Press, Totowa, New Jersey, pp 173–194

94. Mamounas LA, Blue ME, Siuciak JA, Altar CA (1995) Brain-derived neurotrophic factor promotes the survival and sprouting of serotonergic axons in rat brain. J Neurosci 15:7929–7939

95. Tong L, Perez-Polo R (1998) Brain-derived neurotrophic factor (BDNF) protects cultured rat cerebellar granule neurons against glucose deprivation-induced apoptosis. J Neural Transm 105:905–914

96. Zuccato C, Cattaneo E (2009) Brain-derived neurotrophic factor in neurodegenerative diseases. Nat Rev Neurol 5:311–322

97. Altar CA (1999) Neurotrophins and depression. Trends Pharmacol Sci 20:59–61

98. Duman RS, Heninger GR, Nestler EJ (1997) A molecular and cellular theory of depression. Arch Gen Psychiatry 54:597–606

99. Lodge D et al (2002) Ionotropic glutamate receptors as therapeutic targets. F.P. Graham Publishing Co., Johnson city, TN, Biomedical Book Series

100. Pliakas AM, Carlson RR, Neve RL, Konradi C, Nestler EJ, Carlezon WA (2001) Altered responsiveness to cocaine and increased immobility in the forced swim test associated with elevated cAMP response element-binding protein expression in nucleus accumbens. J Neurosci 21:7397–7403

101. Brandoli C, Sanna A, De Bernardi MA, Follesa P, Brooker G, Mocchetti I (1998) Brain-derived neurotrophic factor and basic fibroblast growth factor downregulate NMDA receptor function in cerebellar granule cells. J Neurosci 18:7953–7961

102. Dybala M, Siwek A, Poleszak E, Pilc A, Nowak G (2008) Lack of NMDA–AMPA interaction in antidepressant-like effect of CGP 37849, an antagonist of NMDA receptor, in the forced swim test. J Neural Transm 115:1519–1520

103. Stahl SM (2000) Placebo-controlled comparison of the selective serotonin reuptake inhibitors citalopram and sertraline. Biol Psychiatry 48:894–901

Ionic Glutamate Modulators in Depression (Zinc, Magnesium)

Bernadeta Szewczyk, Ewa Poleszak, Andrzej Pilc, and Gabriel Nowak

Abstract Considerable evidence has accumulated over the past 10 years demonstrating an important role of zinc and magnesium, potent modulators of glutamate receptors, in depression and antidepressant treatment. Clinical reports revealed reduced serum zinc and magnesium in depression, which can be normalized by successful antidepressant treatment. A preliminary clinical study demonstrated the benefit of zinc supplementation in antidepressant therapy in both treatment non-resistant and resistant patients. The clinical efficacy of magnesium treatment was observed in major depression and depressed elderly diabetics with hypomagnesemia. Preclinical studies demonstrated antidepressant activity of zinc and magnesium in a variety of rodent tests and models of depression and suggest a causative role for zinc and magnesium deficiency in the induction of depressive-like symptoms in rodents. This chapter provides an overview of the clinical and experimental evidence that implicates zinc and magnesium in the pathophysiology and therapy of depression in the context of the glutamate hypothesis of this disease.

1 Zinc

1.1 Physiological Functions of Zinc

Zinc is one of the most abundant trace elements in the human body. It is a key structural component of a great number of proteins and a cofactor of enzymes regulating a variety of cellular processes and cellular signaling pathways [1]. In the

G. Nowak (✉)
Department of Neurobiology, Institute of Pharmacology, Polish Academy of Sciences, Smętna 12, PL 31-343 Kraków, Poland
Department of Cytobiology, Collegium Medicum, Jagiellonian University, Medyczna 9, PL 30-688 Kraków, Poland
e-mail: nowak@if-pan.krakow.pl

P. Skolnick (ed.), *Glutamate-based Therapies for Psychiatric Disorders*, Milestones in Drug Therapy, DOI 10.1007/978-3-0346-0241-9_2, © Springer Basel AG 2010

mammalian brain, besides the zinc associated with proteins (95%), there is a specific pool of zinc localized within the synaptic vesicles. Neurons that contain zinc ions in the vesicles of their presynaptic boutons and are present in the cortex, amygdala, and hippocampus are mostly glutamatergic and have been termed gluzinergic neurons [2, 3]. Neurons with terminals containing zinc and that are located in the other brain regions are generally called zinc-enriched neurons (ZEN) [2, 3]. In the spinal cord, the majority of ZEN are GABAergic, and the others are glycinergic [4]. The hippocampus, amygdala, and cortex are the brain regions where the highest zinc concentrations are found [3]. Zinc penetrates the brain via the brain barrier systems (the blood–brain and blood–cerebrospinal fluid barrier) [5]. The process of zinc uptake from extracellular fluids into neurons and glial cells is, in part, regulated by Zip family transporters and the efflux from neurons or uptake to the vesicles is regulated by ZnT family transporters [5–7]. These proteins are encoded by two solute-linked carrier (SLC) gene families: *ZnT* (*SLC30*) and *Zip* (*SLC39*). They exhibit opposite roles in cellular zinc homeostasis, tissue-specific expression, and differential responsiveness to both zinc excess and deficiency [8]. Synaptically released zinc is involved in the modulation of glutamatergic (via both inotropic – ligand-gated ion channels and metabotropic – G-protein linked glutamate receptors, mGluRs), GABAergic, and glycinergic synaptic transmission [6, 9, 10]. The best characterized effect of synaptic zinc is the inhibition of N-methyl-D-aspartate (NMDA) receptors. On the NMDA receptor channel complex, two different mechanisms of action for zinc have been identified: a voltage-independent, noncompetitive (allosteric) inhibition, responsible for reducing channel-opening frequency, and voltage-dependent inhibition, representing an open channel blocking effect [11, 12]. The comparison of NR1/NR2A and NR1/NR2B receptors revealed that the voltage-dependent inhibition is similar in both types of receptors, but the voltage-independent zinc inhibition is highly subunit-specific, with an affinity ranging from low nM for NR1/NR2A receptors to 1 μM for NR1/NR2B receptors and ≥ 10 μM for NR1/NR2C and NR1/NR2D receptors [13, 14]. However, the maximal effect of zinc is smaller at receptors containing NR2A than the NR2B subunit because zinc exerts only a partial inhibition of the NR2A containing receptors yet fully inhibits the NR2B containing receptors [15]. Zinc also modulates α-amino-3-hydroxyl-5-methyl-4-isoxazole-propionate (AMPA) glutamate receptors, although modulation of these receptors differs from that of the NMDA receptors. First, zinc acts as an enhancer of AMPA receptor function and, second, inhibition can be observed only at very high concentrations (in the mM range), and finally the presence of the GluR3 subunit seems to be necessary for zinc modulation [12, 16–18].

1.2   Zinc and Depression

1.2.1   Human Study (Table 1)

The first clinical findings published by Hansen et al. [19] indicated low serum zinc levels in treatment resistant depressed patients. Low serum zinc level was later

Table 1 Summary of clinical and preclinical evidence supporting the involvement of zinc in depression

Clinical evidence	References
Alterations in serum zinc concentrations in depression	
Lower serum zinc level in patients with unipolar depression, postpartum depression, and treatment resistant depression	[20–24]
Negative correlation between serum zinc level and severity of illness in unipolar and postpartum but not in treatment resistant depression	[21–24]
Normalization of serum zinc level after successful antidepressant therapy	[20, 24, 28]
Antidepressant supplementation with zinc in depressed patients	
Reduced depression scores (HDRS, BDI) in patients with unipolar depression treated with clomipramine amitryptiline, citalopram, fluoxetine	[29]
Reduced depression scores (HDRS, BDI, CGI, MADRS) and facilitated the treatment outcome (imipramine) in antidepressant treatment resistant patients	[30]

Preclinical evidence	
Direct antidepressant effect of zinc	
Forced swim test (FST): mice and rats; acute and chronic zinc treatment	[32–36]
Tail suspension test (TST): mice; acute zinc treatment	[35, 37]
Olfactory bulbectomy model (OB): rats; acute and chronic treatment	[34]
Chronic mild stress (CMS): rats; chronic treatment	[43]
Chronic unpredictable stress (CUS): rats; chronic treatment	[45]
Zinc potentiation of the action of subeffective doses of antidepressants	
FST (imipramine, citalopram, fluoxetine)	[32, 35, 47, 48]
TST (imipramine, desipramine, citalopram, paroxetine, bupropion)	[37]
CUS (imipramine)	[45]
Effect of zinc deficiency	
Enhanced depressive-like behavior in the FST and TST in mice and rats	[49–51]
Appearance of anhedonia, anxiety, and anorexia in zinc-deficient rats	[49, 50, 53]
Alterations in serum and brain zinc concentrations after chronic treatment with antidepressants	
Increased serum and hippocampal zinc level after citalopram treatment	[59]
Increased hippocampal and unchanged serum zinc level after imipramine and ECS	[59]
Alteration in synaptic zinc concentration in the rodent brain	
Increased presynaptic zinc level in the hippocampus after chronic ECS and zinc treatment	[54–57]
Increased presynaptic zinc level in the prefrontal cortex after citalopram, imipramine, or zinc treatment	[unpublished]
Increased extracellular zinc level in the prefrontal cortex after citalopram, imipramine, or zinc treatment	[unpublished]

Abbreviations: *HDRS* Hamilton Depression Rating Score; *BDI* Back Depression Inventory Score; *CGI* Clinical Global Impression Scale Score; *MADRS* Montgomery–Asberg Depression Rating Scale Scores; *ECS* electroconvulsive shock

found in major depressed [20–22] and minor depressed subjects (that is, with dysthymic disorder and adjustment disorder with depressed mood) [21]. A lower serum zinc concentration may also accompany antepartum and postpartum depressive symptoms [23]. In both groups of patients (with major and postpartum depression) a significant negative correlation was observed between serum zinc concentration and the severity of depression [21, 23, 24]. There is now some evidence indicating the relationship between activation of the inflammatory response system (IRS) and depression [25, 26]. In turn, IRS activation is accompanied by a decrease in serum zinc level [27]. In fact, in patients with major depression, a low zinc serum level correlated with an increase in the activation of markers of the immune system [20, 21]. Thus, these findings raise the hypothesis that the lower serum zinc observed in depressed patients may, in part, result from a depression-related alteration in the immune-inflammatory system. The other data supporting an important role of zinc in depression comes from the findings that the lower serum zinc level observed in depressed patients could be normalized by successful antidepressant therapy [20, 24, 28]. It was also found that zinc supplementation may enhance standard antidepressant therapy. Nowak et al. [29] described the effect of zinc supplementation on a group of patients with unipolar depression (assessed by the Hamilton Depression Rating Scale [HDRS] and Beck Depression Inventory [BDI]) treated with standard antidepressant therapy such as tricyclic antidepressants and selective serotonin reuptake inhibitors. The analysis of the HDRS and BDI scores after 6 and 12 weeks of treatment revealed that patients who received zinc supplementation of antidepressant treatment display much lower scores than patients treated with placebos and antidepressants. Recently, a beneficial effect of zinc supplementation was found in treatment resistant patients [30]. In this placebo-controlled, randomized double blind study, zinc supplementation augments the efficacy (reduced depression scores measured by Clinical Global Impression [CGI], Montgomery–Asberg Depression Rating Scale [MADRS], BDI, and HDRS) and the speed of the onset of the therapeutic response to imipramine in treatment resistant patients. This data suggests the involvement of zinc/glutamatergic transmission in the psychopathology of drug resistance. Studies conducted in suicide victims and psychiatrically normal controls did not show any difference in the zinc concentration either in the hippocampus or in the frontal cortex, although it was observed that there was a statistically significant decrease in the ability of zinc to inhibit the $[^3H]MK-801$ binding to NMDA in the hippocampus of suicide victims when compared with control subjects [31]. This data revealed that the alterations in the interaction between zinc and NMDA may be involved in the psychopathology underlying suicidality.

1.2.2 Preclinical Studies (Table 1)

The majority of findings that implicated either the antidepressant-like activity of zinc or involvement of zinc in the mechanism of action of antidepressant drugs come from preclinical studies. Zinc exerts antidepressant-like effects in both animal drug screening tests and models of depression. Zinc administered acutely and repeatedly intraperitoneally (*i.p.*) was found to be active (reduced the immobility time) in the forced swimming test (FST) both in mice and rats [32–36]. Zinc was also active in the tail suspension test (TST) in mice both *i.p.* [35] and orally (*p.o.*) [37]. Both of these tests, especially the FST, have a good value for predicting the antidepressant-efficacy of new compounds [38]. Moreover, the antidepressant-like activity of zinc was demonstrated in different models of depression. One of the most validated animal models of depression is the olfactory bulbectomy (OB) model [39]. Removal of the olfactory bulbs results in a number of neurochemical and behavioral changes such as increased hyperactivity and exploratory behavior and significant impairment of learning- and memory-related behavior in a passive-avoidance test [40]. It was found that both acute and chronic administration of zinc reduced the number of trials needed for passive-avoidance learning and OB-induced hyperactivity in rats [34]. Thus, zinc exhibits rapid (acute) antidepressant-like activity in this model, which was an effect observed only for specific serotonin reuptake inhibitors [41]. The rapid antidepressant-like effect of zinc was also observed in the chronic mild stress (CMS) model of depression. In this model, rodents are exposed sequentially to a variety of mild stressors, changed every few hours over a period of weeks or months, which results in a substantial and long-lasting decrease in the responsiveness to rewarding stimuli [42]. The antidepressant-like effect of zinc in the CMS was already seen after 1 week of treatment [43]. By comparison, the significant effect of escitalopram, citalopram (CIT), fluoxetine, or imipramine (IMI) in this test was achieved after 1, 2, 3, and 4 weeks, respectively [44]. The other study indicated that zinc, similar to antidepressants, protects the rats against the depressive-like behavior induced by chronic unpredictable stress (CUS) [45]. It was found that in rats subjected to the CUS procedure, footshock-induced fighting behavior is significantly reduced and chronic treatment with antidepressants may prevent these stress-induced behavioral deficits [46]. Additionally, zinc supplementation was found to enhance the effect of imipramine in this behavioral model of depression [45]. The synergistic effects of zinc and other antidepressants were also shown in the FST and TST [33, 35, 47, 48].

Data published recently revealed a causative role of zinc deficiency in the induction of depressive-like symptoms. Mice and rats treated with a zinc-deficient diet demonstrated depressive-like behavior (increase in the immobility time) in the FST and TST [49–52]. These behavioral disturbances observed in mice were normalized after chronic desipramine treatment [50]. Additionally, zinc-deficient mice exhibit anxiety-related behavior, as observed in the novelty suppressed feeding test and measured as increased latencies to eat [50]. This effect was reversed by chronic treatment with desipramine and Hypericum perforatum (treatment effective in atypical depression) [50]. The other study performed in rats showed that dietary zinc deficiency leads to anhedonia, anxiety, and anorexia, which are the common comorbid symptoms of depression [49, 53]. These data suggest that experimentally induced zinc-deficiency might be used as an alternative to model certain aspects of depression.

The important role of zinc in the mechanism of antidepressants was also supported by several biochemical and immunohistochemical evidences. Studies in which Timm's histochemical method (a method imaging the presynaptic-vesicle pool of zinc) was used showed that chronic electroconvulsive shock (ECS) but not fluoxetine or desipramine treatment increases the presynaptic/vesicular zinc level in rat hippocampus [54–56]. We found recently (using a modified Timm's method) that chronic treatment of citalopram and imipramine increases the pool of presynaptic zinc in the rat prefrontal cortex but not in the hippocampus (unpublished data). For comparison, chronic treatment of zinc increases the pool of synaptic zinc in both the rat hippocampus [57] and prefrontal cortex (unpublished data). The results published by Opoka et al. [58] showed for the first time that acute *i.p.* zinc administration increases cortical extracellular zinc pool (determined by the anodic stripping voltammetric method), which indicates the fast brain penetration of zinc and may explain its rapid pharmacological effects observed in the animal tests and models of depression. Likewise, our unpublished data showed an increase of the extracellular zinc level in the prefrontal cortex but not in the hippocampus after chronic citalopram and imipramine treatment. Previously, we showed that chronic treatment with imipramine and citalopram induced a slight increase in the total level of zinc in the rat hippocampus (determined by flame atomic absorption spectrometry) and a slight decrease in the rat neocortex [59]. In turn, chronic ECS treatment induced a robust increase in the total zinc level in both brain regions [59]. The data presented above suggests an "ECS-like" profile of antidepressant action induced by zinc treatment.

The main role of zinc in the central nervous system is the inhibition of the NMDA receptor complex [60]. Chronic treatment with imipramine increases the potency of zinc to inhibit [^{3}H]MK-801 binding to NMDA in the mouse cortex but not the hippocampus [61], which may be associated with the existence of multiple forms of the NMDA receptor complex (region specific subunit composition and different pharmacological properties) [60]. This data and the other presented above clearly suggest the involvement of zinc in the mechanism of action of antidepressant drugs.

1.3   Mechanism of Antidepressant Activity of Zinc

1.3.1   Involvement of the Glutamate System

One of the best established mechanisms involved in the antidepressant-like activity of zinc is the inhibitory modulation of glutamate signaling. There is some direct evidence indicating the involvement of the NMDA receptor complex in the antidepressant-like activity of zinc in the FST. In fact, N-methyl-D-aspartic acid (NMDA) administration antagonized the effect induced by zinc treatment in the FST in both rats and mice [62]. The antidepressant-like effect of zinc observed in the FST was also abolished by D-serine cotreatment, which is an agonist of the glycine$_B$ site of the NMDA receptor complex [63]. Moreover, the joint administration of NMDA antagonists (CGP-37849, L-701,324, MK-801, D-cycloserine) and zinc in low doses, which were ineffective in FST produced a significant reduction of the immobility time in this test [62]. Receptor binding and electrophysiological experiments showed that chronic zinc administration reduced the affinity of glycine to glycine/NMDA receptors and NMDA receptor reactivity, respectively, in the rat

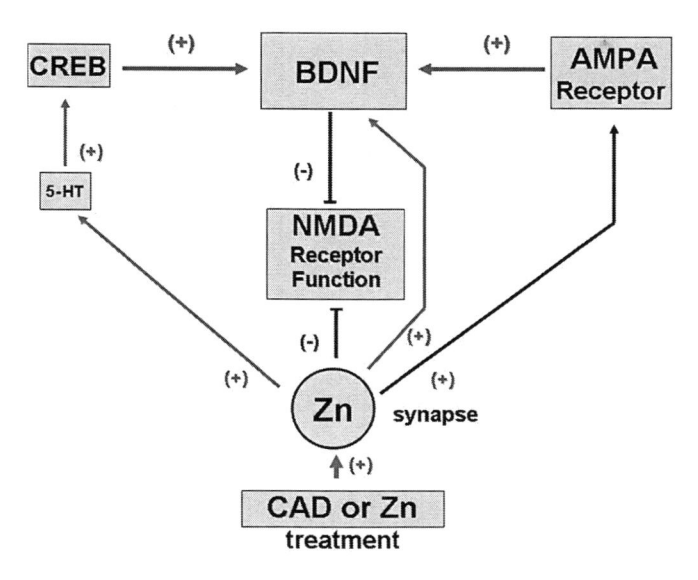

Fig. 1 A simplistic representation of the mechanism linked to antidepressant-induced increases in synaptic zinc (Zn) concentration. The depiction is based on our unpublished data and previous reports [62, 64, 65]. Conventional antidepressants (CAD) or zinc treatment increases the concentration of zinc in the synapse. Synaptic zinc antagonizes the NMDA receptor complex directly or by enhancing AMPA receptors and/or serotonin (5-HT) pathways involving CREB/BDNF pathway

frontal cortex [64, 65]. Furthermore, it is suggested that the antidepressant activity of zinc may involve the NMDA-nitric oxide pathway. In fact, zinc is an inhibitor of nitric oxide synthase (NOS) activity [66]. Rosa et al. [35] demonstrated that the antidepressant-like action of zinc in the FST was prevented by pretreatment with L-arginine, which indicates the zinc-induced inhibition of NOS activity in this test. The antidepressant effect of zinc observed in the FST seems to be also associated with the AMPA receptor modulation. Administration of the NBQX, an AMPA receptor antagonist, abolished the antidepressant-like effects elicited by zinc in the FST in mice. Moreover, low, ineffective doses of zinc and the AMPA receptor potentiator CX614, administered jointly, exhibit a significant reduction of the immobility time in the FST [62] (Fig. 1).

1.3.2 Involvement of the Serotonergic System

An important role of the serotonergic system in the antidepressant-like effect of zinc was also demonstrated. In fact, the synergistic effect of zinc with "serotonergic" antidepressants, such as serotonin, reuptake inhibitors (citalopram, fluoxetine), and dual serotonin and noradrenaline reuptake inhibitor (imipramine), were shown in the FST [33, 35, 47, 48]. A similar effect was found for fluoxetine, paroxetine, and imipramine in the TST [37]. Moreover, pretreatment with pCPA (an inhibitor of serotonin synthesis), WAY 1006335 (5-HT_{1A} receptor antagonist), and ritanserin (5-$HT_{2A/C}$ receptor antagonist) completely reduced the antidepressant-like action induced by zinc in the FST [48]. On the other hand, chronic zinc treatment increases the density of 5-HT_{1A} receptors in the rat hippocampus and 5-HT_{2A} receptors in the rat frontal cortex [65] (Fig. 1).

1.3.3   Involvement of BDNF

A pivotal role of the brain-derived neurotrophic factor (BDNF) in the pathophysiology and therapy of depression has been already established [67, 68]. Reports published recently suggested that regulation of mRNA and protein levels of BDNF may be also involved in the mechanism of zinc antidepressant action. It was found that chronic high dose zinc treatment increases BDNF mRNA and protein level in the rat cortex [36, 65, 69], while a low dose increases BDNF mRNA and protein level in the hippocampus [43]. These effects suggest that mechanisms regulating the BDNF levels are more sensitive to zinc in the hippocampus than in the cortex (Fig. 1).

1.3.4   Involvement of the GSK-3 Enzyme

Another possible mechanism involved in the antidepressant-like activity of zinc is the inhibition of the glycogen synthase kinase-3 (GSK-3) enzyme. GSK-3 is a serine/threonine kinase involved in glycogen metabolism [70]. Recent data indicated that either antidepressants or ECS inhibit GSK-3 activity and GSK-3 inhibitors induce antidepressant-like effects in the FST [71, 72]. Since zinc was also found to inhibit GSK-3 activity [73], we hypothesize that the antidepressant activity of zinc may be in part mediated by the modulation of GSK-3. Additionally, it was found that inhibition of GSK-3 enhances cyclic AMP response element binding protein (CREB), which in turn regulates BDNF activity. Thus, zinc can influence the BDNF function via the inhibition of GSK-3 [74].

2   Magnesium

2.1   Physiological Functions of Magnesium

Magnesium is an essential cation, involved in a broad variety of physiological processes. In a healthy human, magnesium is mainly distributed between bones (60%) and soft tissues: muscles, heart, and liver (40%). In tissue, the intracellular magnesium fraction is mostly bound to chelators (adenosine triphosphate ATP, adenosine diphosphate ADP), proteins, RNA, DNA, phospholipids, and citrate [75, 76]. Only 2–3% of intracellular magnesium accounts for the free pool, although this free fraction is critical for regulating the intracellular magnesium homeostasis and cellular function [75–77].

About 1% of the total body magnesium is localized extracellularly, mainly in blood (serum and red blood cells), where it is present in three fractions: protein bound (19%), complexed to anions such as citrate, phosphate, and bicarbonate (14%), and ionized (biologically active form, 67%) [76, 78]. The balance between cerebrospinal fluid (CSF) magnesium concentration and plasma magnesium concentration is regulated by the active transport between these two compartments [79]. This mechanism leads to the stabilization of the intracerebral magnesium concentrations even in the instance of magnesium depletion [80].

Magnesium is a cofactor for hundreds of enzymes involved in numerous metabolically important reactions, especially those involving ATP [76, 81]. It is necessary for protein and nucleic acid synthesis, regulation of the cell cycle, control of mitochondrial processes, membrane stability, and cytoskeletal integrity [76, 82]. Magnesium plays an important role in oxidative phosphorylation processes and the regulation of kinases and Ca^{2+} signaling [81, 83]. It is also a potent antagonist of the NMDA receptor complex [79]. The activation of the NMDA receptor ion channel is blocked by magnesium in a voltage-dependent manner, and this blockade occurs when the concentration of magnesium is less than 1 mM, which is within the range of the magnesium level found in CSF and plasma [79, 84]. Lowering extracellular magnesium concentration was found to increase central hyperexcitability due to the disinhibition of the NMDA receptor channel [84]. Recent data from experimental and clinical studies suggests the important role of magnesium deficiency in many diseases. Generally, hypomagnesemia manifests as cardiac, neuromuscular, and neurological disorders [76].

2.2 Magnesium and Depression

2.2.1 Human Study (Table 2)

The involvement of magnesium in pathophysiology and the treatment of depression and other psychiatric diseases have been suspected for decades. The first data that indicated the beneficial effects of magnesium in the therapy of depression was published in 1921 by Weston [85], who found magnesium sulfate effective in the treatment of agitated depression. The next years brought new findings indicating disturbances of magnesium in blood and cerebrospinal fluid's (CSF) concentration in depression, although consistent results have not been obtained. Some clinical data linked magnesium deficiency with major depression and related or accompanying depression and mental health problems. A lower total serum/plasma or erythrocyte magnesium level was found in patients with major depression [86–90] and in older depressed patients with diabetes [91]. In some of these studies, a correlation between the serum magnesium level and incidence of depressive symptoms was observed [86–88, 90]. A low magnesium level was also found in the CSF of depressed patients who had made suicide attempts [92]. On the other hand, no alterations or increase in the serum magnesium concentration have also been observed [93–97].

Other study supporting a role for magnesium in depressive disorder include report that treatment of sertraline and amitryptiline increases magnesium levels in erythrocytes [90]. The increased magnesium level was also found in depressed patients responding to lithium [98]. Clinical efficacy of magnesium treatment was observed in patients with major depression [99] and in depressed elderly diabetics with hypomagnesemia [100] as well as in mania [101], rapid cycling bipolar disorder [102], and fatigue syndrome [103], disorders which might be related to or accompany depression. In addition, it was found that supplementing lithium, benzodiazepines, and neuropleptics with magnesium significantly reduced the effective doses of these drugs [104].

2.2.2 Preclinical Studies (Table 2)

Most of the evidence suggesting a relationship of magnesium and depression comes from preclinical studies. Mice fed with a magnesium-deficient diet displayed

Table 2 Summary of preclinical and clinical evidence supporting the involvement of magnesium in depression

Clinical evidence	References
Alterations in serum magnesium concentrations in depression	
Low magnesium level in depressed patients who had made suicide attempts (cerebrospinal fluid)	[92]
Low magnesium level in patients with major depression and depressed patients with diabetes (serum)	[86–91]
Increased or unchanged magnesium level in patients with major depression	[93–97]
Correlation between serum magnesium level and incidence of depressive symptoms	[86–88, 90]
Increased magnesium level in patients treated with sertraline, amitryptyline, and lithium	[90, 98]
Clinical efficacy of magnesium treatment	
Major depression	[98]
Depressed elderly diabetics with hypomagnesemia	[91]
Disorders related or accompanying depression (mania, rapid cycling bipolar disorder, fatigue syndrome)	[101–103]
Preclinical evidence	
Direct antidepressant effect of magnesium	
Forced swim test (FST): mice and rats; acute and chronic magnesium treatment	[108–112]
Effect of magnesium deficiency	
Enhanced depression-like behavior in FST and increased anxiety-like behavior in the light/dark and open field test	[105]
Magnesium potentiation of the action of subeffective doses of antidepressants	
FST (imipramine, fluoxetine, citalopram, tianeptine, bupropion)	[110, 112, 116]

enhanced depression-like behavior in the FST and increased anxiety-related behavior in the light/dark and open field test [105]. Moreover, depression- and anxiety-related disturbances observed in the behavior of mice produced by magnesium depletion was reversed by antidepressant and anxiolytic drugs, respectively [105]. It was also found that magnesium depletion in mice leads to a reduction in offensive and an increase in defensive behavior [106]. Furthermore, the correlation between the erythrocyte magnesium level and behavior in mice was also found. Mice with a low erythrocyte magnesium concentration exhibit more restlessness and more aggressive behavior under stressful conditions than mice with a high erythrocyte magnesium level [107]. An antidepressant-like effect of magnesium was observed in the FST in rodents. Magnesium treatment reduces the immobility time in the FST in both mice and rats [108–112] and potentiates the action of subeffective doses of antidepressant drugs [111–113]. Magnesium also exhibits anxiolytic-like activity, observed in the elevated plus-maze test as an increase in the number of open arm entries [109], and enhances the anxiolytic-like effects of classical benzodiazepines in this test [114].

2.3   Mechanism of Antidepressant Activity of Magnesium

2.3.1   Involvement of the Glutamate System

Magnesium is a potent antagonist of the NMDA receptor complex [115], so it is highly possible that the antidepressant action of magnesium is induced via this receptor complex. In fact, magnesium induced antidepressant-like activity observed in the FST was antagonized by NMDA [111] and D-serine cotreatment [114]. On the other hand, subactive in the FST doses of magnesium were potentiated by subactive doses of the NMDA receptor complex antagonists (CGP 37849, L-701,324, D-cycloserine, MK-801) [111].

2.3.2   Involvement of Serotonergic System

The involvement of the serotonergic system in the antidepressant-like effect of magnesium in the FST was suggested recently. The enhancement of antidepressant-like activity by the joint administration of subeffective doses of magnesium salts and citalopram or fluoxetine (serotonin reuptake inhibitors), imipramine (mixed serotonin, noradrenaline reuptake inhibitor), and tianeptine (enhancer of serotonin reuptake) was observed in the FST in mice [112, 113, 116]. Also, a reduced antidepressant-like activity of magnesium in the FST after a depletion of serotonin by pCPA (an inhibitor of serotonin synthesis) was demonstrated [116]. The diminished antidepressant-like activity of magnesium in the FST was also observed after pretreatment with NAN-190 ($5-HT_{1A}$ receptor antagonist), WAY 100635 (selective $5-HT_{2A}$ receptor antagonist), ritanserin ($5-HT_{2A/C}$ receptor antagonist), and ketanserin (a preferential $5-HT_{2A}$ receptor antagonist) further confirming the contribution of the serotonergic system in the antidepressant-like action of magnesium [112, 116]. In addition, a direct enhancing effect of magnesium on the $5HT_{1A}$ serotonin receptor transmission was reported [117].

2.3.3 Involvement of the Catecholaminergic System

Recent data, published by Cardoso et al. [112], indicated that the antidepressant-like effect of magnesium in the FST may depend on its interaction with noradrenergic and dopaminergic systems. They found that pretreatment of mice with prazosine (α_1-receptor antagonist), yohimbine (α_2-receptor antagonist), haloperidol (nonselective dopaminergic receptor antagonist), SCH23390 (dopamine D_1 receptor antagonist), and sulpiride (dopamine D_2 receptor antagonist) reduced the antidepressant-like action induced by magnesium in the FST [112]. On the contrary, our unpublished data indicates the lack of influence of the selective noradrenergic neurotoxin N-(2-chloroethyl)-N-ethyl-2-bromobenzylamine (DSP-4) pretreatment on magnesium antidepressant-like activity in the FST.

2.3.4 Involvement of the GSK-3 Enzyme

The other possible target of the antidepressant-like activity of magnesium might be a GSK-3 enzyme. As was mentioned above, GSK-3 inhibitors exhibit antidepressant-like effects in the FST in mice [71, 72] and antidepressant drugs or ECS inhibit the GSK-3 phosphorylation activity. Magnesium like zinc and lithium is also a potent inhibitor of this enzyme [72].

3 Conclusions

Zinc and magnesium, endogenous modulators of glutamate receptors, exhibit antidepressant activity as well as being involved in the mechanism(s) of depression and antidepressant therapy. Although the preclinical data is very convincing, there is a need for additional clinical studies determining efficacy of these ions in depressive disorders. Since zinc and magnesium are already registered as diet supplements and widely used, it would be relatively easy and inexpensive to include them in antidepressant treatment as adjunctive agents.

References

1. Takeda A (2000) Movement of zinc and its functional significance in the brain. Brain Res Brain Res Rev 34:137–148
2. Frederickson CJ, Moncrieff DW (1994) Zinc-containing neurons. Biol Signals 3:127–139

3. Frederickson CJ, Suh SW, Silva D, Frederickson CJ, Thompson RB (2000) Importance of zinc in the central nervous system: the zinc-containing neuron. J Nutr 130:1471S–1483S
4. Wang Z, Li JY, Dahlstrom A, Danscher G (2001) Zinc-enriched GABAergic terminals in mouse spinal cord. Brain Res 921:165–172
5. Takeda A, Tamano H (2009) Insight into zinc signaling from dietary zinc deficiency. Brain Res Rev 62:33–44
6. Law W, Kelland EE, Sharp P, Toms NJ (2003) Characterisation of zinc uptake into rat cultured cerebrocortical oligodendrocyte progenitor cells. Neurosci Lett 352:113–116
7. Seve M, Chimienti F, Devergnas S, Favier A (2004) In silico identification and expression of SLC30 family genes: an expressed sequence tag data mining strategy for the characterization of zinc transporters' tissue expression. BMC Genomics 5:32
8. Liuzzi JP, Cousins RJ (2004) Mammalian zinc transporters. Annu Rev Nutr 24:151–172
9. Laube B (2002) Potentiation of inhibitory glycinergic neurotransmission by Zn2+: a synergistic interplay between presynaptic P2X2 and postsynaptic glycine receptors. Eur J Neurosci 16:1025–1036
10. Mocchegiani E, Bertoni-Freddari C, Marcellini F, Malavolta M (2005) Brain, aging and neurodegeneration: role of zinc ion availability. Prog Neurobiol 75:367–390
11. Christine CW, Choi DW (1990) Effect of zinc on NMDA receptor-mediated channel currents in cortical neurons. J Neurosci 10:108–116
12. Paoletti P, Vergnano AM, Barbour B, Casado M (2009) Zinc at glutamatergic synapses. Neuroscience 158:126–136
13. Chen N, Moshaver A, Raymond LA (1997) Differential sensitivity of recombinant N-methyl-D-aspartate receptor subtypes to zinc inhibition. Mol Pharmacol 51:1015–1023
14. Paoletti P, Ascher P, Neyton J (1997) High-affinity zinc inhibition of NMDA NR1-NR2A receptors. J Neurosci 17:5711–5725
15. Williams K (1996) Separating dual effects of zinc at recombinant N-methyl-D-aspartate receptors. Neurosci Lett 215:9–12
16. Mayer ML, Vyklicky L Jr, Westbrook GL (1989) Modulation of excitatory amino acid receptors by group IIB metal cations in cultured mouse hippocampal neurones. J Physiol 415:329–350
17. Rassendren FA, Lory P, Pin JP, Nargeot J (1990) Zinc has opposite effects on NMDA and non-NMDA receptors expressed in Xenopus oocytes. Neuron 4:733–740
18. Dreixler JC, Leonard JP (1994) Subunit-specific enhancement of glutamate receptor response by zinc. Brain Res Mol Brain Res 22:144–150
19. Hansen CR Jr, Malecha M, Mackenzie TB, Kroll J (1983) Copper and zinc deficiencies in association with depression and neurological findings. Biol Psychiatry 18:395–401
20. McLoughlin IJ, Hodge JS (1990) Zinc in depressive disorder. Acta Psychiatr Scand 82:451–453
21. Maes M, D'Haese PC, Scharpe S, D'Hondt P, Cosyns P, De Broe ME (1994) Hypozincemia in depression. J Affect Disord 31:135–140
22. Nowak G, Zieba A, Dudek D, Krosniak M, Szymaczek M, Schlegel-Zawadzka M (1999) Serum trace elements in animal models and human depression. Part I. Zinc. Hum Psychopharmacol Clin Exp 14:83–86
23. Wojcik J, Dudek D, Schlegel-Zawadzka M, Grabowska M, Marcinek A, Florek E, Piekoszewski W, Nowak RJ, Opoka W, Nowak G (2006) Antepartum/postpartum depressive symptoms and serum zinc and magnesium levels. Pharmacol Rep 58:571–576
24. Schlegel_Zawadzka M, Zieba A, Dudek D, Krosniak M, Szymaczek M, Nowak G (2000) Effect of depression and of antidepressant therapy on serum zinc levels – a preliminary clinical study. In: Roussel AM, Anderson RA, Favrier AE (eds) Trace elements in man and animals 10. Kluwer Academic Plenum Press, New York, pp 607–610
25. Anisman H, Merali Z, Poulter MO, Hayley S (2005) Cytokines as a precipitant of depressive illness: animal and human studies. Curr Pharm Des 11:963–972
26. Maes M, Yirmyia R, Noraberg J, Brene S, Hibbelen J, Perini G, Kubera M, Bob P, Lerer B, Maj M (2009) The inflammatory & neurodegenerative (I&ND) hypothesis of depression:

leads for future research and new drug developments in depression. Metab Brain Dis 24:27–53

27. Srinivas U, Braconier JH, Jeppsson B, Abdulla M, Akesson B, Ockerman PA (1988) Trace element alterations in infectious diseases. Scand J Clin Lab Invest 48:495–500

28. Maes M, Vandoolaeghe E, Neels H, Demedts P, Wauters A, Meltzer HY, Altamura C, Desnyder R (1997) Lower serum zinc in major depression is a sensitive marker of treatment resistance and of the immune/inflammatory response in that illness. Biol Psychiatry 42:349–358

29. Nowak G, Siwek M, Dudek D, Zieba A, Pilc A (2003) Effect of zinc supplementation on antidepressant therapy in unipolar depression: a preliminary placebo-controlled study. Pol J Pharmacol 55:1143–1147

30. Siwek M, Dudek D, Paul IA, Sowa-Kucma M, Zieba A, Popik P, Pilc A, Nowak G (2009) Zinc supplementation augments efficacy of imipramine in treatment resistant patients: a double blind, placebo-controlled study. J Affect Disord 118:187–195

31. Nowak G, Szewczyk B, Sadlik K, Piekoszewski W, Trela F, Florek E, Pilc A (2003) Reduced potency of zinc to interact with NMDA receptors in hippocampal tissue of suicide victims. Pol J Pharmacol 55:455–459

32. Kroczka B, Branski P, Palucha A, Pilc A, Nowak G (2001) Antidepressant-like properties of zinc in rodent forced swim test. Brain Res Bull 55:297–300

33. Kroczka B, Zieba A, Dudek D, Pilc A, Nowak G (2000) Zinc exhibits an antidepressant-like effect in the forced swimming test in mice. Pol J Pharmacol 52:403–406

34. Nowak G, Szewczyk B, Wieronska JM, Branski P, Palucha A, Pilc A, Sadlik K, Piekoszewski W (2003) Antidepressant-like effects of acute and chronic treatment with zinc in forced swim test and olfactory bulbectomy model in rats. Brain Res Bull 61:159–164

35. Rosa AO, Lin J, Calixto JB, Santos AR, Rodrigues AL (2003) Involvement of NMDA receptors and L-arginine-nitric oxide pathway in the antidepressant-like effects of zinc in mice. Behav Brain Res 144:87–93

36. Franco JL, Posser T, Brocardo PS, Trevisan R, Uliano-Silva M, Gabilan NH, Santos AR, Leal RB, Rodrigues AL, Farina M, Dafre AL (2008) Involvement of glutathione ERK1/2 phosphorylation and BDNF expression in the antidepressant-like effect of zinc in rats. Behav Brain Res 188:316–323

37. Cunha MP, Machado DG, Bettio LE, Capra JC, Rodrigues AL (2008) Interaction of zinc with antidepressants in the tail suspension test. Prog Neuropsychopharmacol Biol Psychiatry 32:1913–1920

38. Nestler EJ, Barrot M, DiLeone RJ, Eisch AJ, Gold SJ, Monteggia LM (2002) Neurobiology of depression. Neuron 34:13–25

39. Kelly JP, Wrynn AS, Leonard BE (1997) The olfactory bulbectomized rat as a model of depression: an update. Pharmacol Ther 74:299–316

40. Song C, Leonard BE (2005) The olfactory bulbectomised rat as a model of depression. Neurosci Biobehav Rev 29:627–647

41. van Riezen H, Leonard BE (1990) Effects of psychotropic drugs on the behavior and neurochemistry of olfactory bulbectomized rats. Pharmacol Ther 47:21–34

42. Papp M, Moryl E, Willner P (1996) Pharmacological validation of the chronic mild stress model of depression. Eur J Pharmacol 296:129–136

43. Sowa-Kucma M, Legutko B, Szewczyk B, Novak K, Znojek P, Poleszak E, Papp M, Pilc A, Nowak G (2008) Antidepressant-like activity of zinc: further behavioral and molecular evidence. J Neural Transm 115:1621–1628

44. Sanchez C, Gruca P, Papp M (2003) R-citalopram counteracts the antidepressant-like effect of escitalopram in a rat chronic mild stress model. Behav Pharmacol 14:465–470

45. Cieslik K, Klenk-Majewska B, Danilczuk Z, Wrobel A, Lupina T, Ossowska G (2007) Influence of zinc supplementation on imipramine effect in a chronic unpredictable stress (CUS) model in rats. Pharmacol Rep 59:46–52

46. Ossowska G, Zebrowska-Lupina I, Danilczuk Z, Klenk-Majewska B (2002) Repeated treatment with selective serotonin reuptake inhibitors but not anxiolytics prevents the stress-induced deficit of fighting behavior. Pol J Pharmacol 54:373–380
47. Szewczyk B, Branski P, Wieronska JM, Palucha A, Pilc A, Nowak G (2002) Interaction of zinc with antidepressants in the forced swimming test in mice. Pol J Pharmacol 54:681–685
48. Szewczyk B, Poleszak E, Wlaz P, Wrobel A, Blicharska E, Cichy A, Dybala M, Siwek A, Pomierny-Chamiolo L, Piotrowska A, Branski P, Pilc A, Nowak G (2009) The involvement of serotonergic system in the antidepressant effect of zinc in the forced swim test. Prog Neuropsychopharmacol Biol Psychiatry 33:323–329
49. Tassabehji NM, Corniola RS, Alshingiti A, Levenson CW (2008) Zinc deficiency induces depression-like symptoms in adult rats. Physiol Behav 95:365–369
50. Whittle N, Lubec G, Singewald N (2009) Zinc deficiency induces enhanced depression-like behaviour and altered limbic activation reversed by antidepressant treatment in mice. Amino Acids 36:147–158
51. Tamano H, Kan F, Kawamura M, Oku N, Takeda A (2009) Behavior in the forced swim test and neurochemical changes in the hippocampus in young rats after 2-week zinc deprivation. Neurochem Int 55:536–541
52. Watanabe M, Tamano H, Kikuchi T, Takeda A (2009) Susceptibility to stress in young rats after 2-week zinc deprivation. Neurochem Int. doi:10.1016/j.neuint.2009.11.014
53. Takeda A, Tamano H, Kan F, Itoh H, Oku N (2007) Anxiety-like behavior of young rats after 2-week zinc deprivation. Behav Brain Res 177:1–6
54. Gombos Z, Spiller A, Cottrell GA, Racine RJ, McIntyre BW (1999) Mossy fiber sprouting induced by repeated electroconvulsive shock seizures. Brain Res 844:28–33
55. Vaidya VA, Siuciak JA, Du F, Duman RS (1999) Hippocampal mossy fiber sprouting induced by chronic electroconvulsive seizures. Neuroscience 89:157–166
56. Lamont SR, Paulls A, Stewart CA (2001) Repeated electroconvulsive stimulation, but not antidepressant drugs, induces mossy fibre sprouting in the rat hippocampus. Brain Res 893:53–58
57. Szewczyk B, Sowa M, Czupryn A, Wieronska JM, Branski P, Sadlik K, Opoka W, Piekoszewski W, Smialowska M, Skangiel-Kramska J, Pilc A, Nowak G (2006) Increase in synaptic hippocampal zinc concentration following chronic but not acute zinc treatment in rats. Brain Res 1090:69–75
58. Opoka W, Sowa-Kucma M, Kowalska M, Bas B, Golembiowska K, Nowak G (2008) Intraperitoneal zinc administration increases extracellular zinc in the rat prefrontal cortex. J Physiol Pharmacol 59:477–487
59. Nowak G, Schlegel-Zawadzka M (1999) Alterations in serum and brain trace element levels after antidepressant treatment. Part I. Zinc. Biol Trace Elem Res 67:85–92
60. Bresink I, Danysz W, Parsons CG, Mutschler E (1995) Different binding affinities of NMDA receptor channel blockers in various brain regions-indication of NMDA receptor heterogeneity. Neuropharmacology 34:533–540
61. Szewczyk B, Kata R, Nowak G (2001) Rise in zinc affinity for the NMDA receptor evoked by chronic imipramine is species-specific. Pol J Pharmacol 53:641–645
62. Szewczyk B, Poleszak E, Sowa-Kucma M, Wrobel A, Slotwinski S, Listos J, Wlaz P, Cichy A, Siwek M, Dybala M, Golembiowska K, Pilc A, Nowak G (2010) The involvement of NMDA and AMPA receptors in the mechanism of antidepressant-like action of zinc in forced swim test. Amino Acids 39:205–217
63. Poleszak E, Szewczyk B, Wlaz A, Fidecka S, Wlaz P, Pilc A, Nowak G (2008) D-serine, a selective glycine/N-methyl-D-aspartate receptor agonist, antagonizes the antidepressant-like effects of magnesium and zinc in mice. Pharmacol Rep 60:996–1000
64. Bobula B, Hess G (2008) Antidepressant treatments-induced modifications of glutamatergic transmission in rat frontal cortex. Pharmacol Rep 60:865–871
65. Cichy A, Sowa-Kucma M, Legutko B, Pomierny-Chamiolo L, Siwek A, Piotrowska A, Szewczyk B, Poleszak E, Pilc A, Nowak G (2009) Zinc-induced adaptive changes in NMDA/glutamatergic and serotonergic receptors. Pharmacol Rep 61:1184–1191

66. Mittal CK, Harrell WB, Mehta CS (1995) Interaction of heavy metal toxicants with brain constitutive nitric oxide synthase. Mol Cell Biochem 149–150:263–265

67. Duman RS (2004) Role of neurothropic factors in the etiology and treatment of mood disorders. Neuromolecular Med 5:11–25

68. Schmidt HD, Duman RS (2007) The role of neurotrophic factors in adult hippocampal neurogenesis, antidepressant treatments and animal models of depressive-like behavior. Behav Pharmacol 18:391–418

69. Nowak G, Legutko B, Szewczyk B, Papp M, Sanak M, Pilc A (2004) Zinc treatment induces cortical brain-derived neurotrophic factor gene expression. Eur J Pharmacol 492:57–59

70. Embi N, Rylatt DB, Cohen P (1980) Glycogen synthase kinase-3 from rabbit skeletal muscle. Separation from cyclic-AMP-dependent protein kinase and phosphorylase kinase. Eur J Biochem 107:519–527

71. Kaidanovich-Beilin O, Milman A, Weizman A, Pick CG, Eldar-Finkelman H (2004) Rapid antidepressive-like activity of specific glycogen synthase kinase-3 inhibitor and its effect on beta-catenin in mouse hippocampus. Biol Psychiatry 55:781–784

72. Gould TD, Manji HK (2005) Glycogen synthase kinase-3: a putative molecular target for lithium mimetic drugs. Neuropsychopharmacology 30:1223–1237

73. Ilouz R, Kaidanovich O, Gurwitz D, Eldar-Finkelman H (2002) Inhibition of glycogen synthase kinase-3beta by bivalent zinc ions: insight into the insulin-mimetic action of zinc. Biochem Biophys Res Commun 295:102–106

74. Szewczyk B, Poleszak E, Sowa-Kucma M, Siwek M, Dudek D, Ryszewska-Pokrasniewicz B, Radziwon-Zaleska M, Opoka W, Czekaj J, Pilc A, Nowak G (2008) Antidepressant activity of zinc and magnesium in view of the current hypothesis of antidepressant action. Pharmacol Rep 60:588–589

75. Murphy E (2000) Mysteries of magnesium homeostasis. Circ Res 86:245–248

76. Touyz RM (2004) Magnesium in clinical medicine. Front Biosci 9:1278–1293

77. Romani AM, Scarpa A (2000) Regulation of cellular magnesium. Front Biosci 5:D720–D734

78. Altura BM (1994) Introduction: importance of Mg in physiology and medicine and the need for ion selective electrodes. Scand J Clin Lab Invest Suppl 217:5–9

79. Morris ME (1992) Brain and CSF magnesium concentrations during magnesium deficit in animals and humans: neurological symptoms. Magnes Res 5:303–313

80. Murck H (2002) Magnesium and affective disorders. Nutr Neurosci 5:375–389

81. Ryan MF (1991) The role of magnesium in clinical biochemistry: an overview. Ann Clin Biochem 28(Pt 1):19–26

82. Wolf FI, Trapani V, Cittadini A (2008) Magnesium and the control of cell proliferation: looking for a needle in a haystack. Magnes Res 21:83–91

83. Yang ZW, Wang J, Zheng T, Altura BT, Altura BM (2000) Low [Mg(2+)](o) induces contraction and [Ca(2+)](i) rises in cerebral arteries: roles of ca(2+), PKC, and PI3. Am J Physiol Heart Circ Physiol 279:H2898–H2907

84. Mayer ML, Westbrook GL, Guthrie PB (1984) Voltage-dependent block by Mg2+ of NMDA responses in spinal cord neurones. Nature 309:261–263

85. Weston PG (1921) Magnesium as a sedative. Am J Psychiatry 278:637–638

86. Frizel D, Coppen A, Marks V (1969) Plasma magnesium and calcium in depression. Br J Psychiatry 115:1375–1377

87. Linder J, Brismar K, Beck-Friis J, Saaf J, Wetterberg L (1989) Calcium and magnesium concentrations in affective disorder: difference between plasma and serum in relation to symptoms. Acta Psychiatr Scand 80:527–537

88. Hashizume N, Mori M (1990) An analysis of hypermagnesemia. Jpn J Med 29:368–372

89. Zieba A, Kata R, Dudek D, Schlegel-Zawadzka M, Nowak G (2000) Serum trace elements in animal models and human depression: Part III. Magnesium. Relationship with copper. Hum Psychopharmacol 15:631–635

90. Nechifor M (2008) Interactions between magnesium and psychotropic drugs. Magnes Res 21:97–100

91. Barragan-Rodriguez L, Rodriguez-Moran M, Guerrero-Romero F (2007) Depressive symptoms and hypomagnesemia in older diabetic subjects. Arch Med Res 38:752–756

92. Banki CM, Vojnik M, Papp Z, Balla KZ, Arato M (1985) Cerebrospinal fluid magnesium and calcium related to amine metabolites, diagnosis, and suicide attempts. Biol Psychiatry 20:163–171

93. Kirov GK, Birch NJ, Steadman P, Ramsey RG (1994) Plasma magnesium levels in a population of psychiatric patients: correlations with symptoms. Neuropsychobiology 30:73–78

94. Young LT, Robb JC, Levitt AJ, Cooke RG, Joffe RT (1996) Serum Mg2+ and Ca2+/Mg2+ ratio in major depressive disorder. Neuropsychobiology 34:26–28

95. Frazer A, Ramsey TA, Swann A, Bowden C, Brunswick D, Garver D, Secunda S (1983) Plasma and erythrocyte electrolytes in affective disorders. J Affect Disord 5:103–113

96. Widmer J, Bovier P, Karege F, Raffin Y, Hilleret H, Gaillard JM, Tissot R (1992) Evolution of blood magnesium, sodium and potassium in depressed patients followed for three months. Neuropsychobiology 26:173–179

97. Widmer J, Henrotte JG, Raffin Y, Bovier P, Hilleret H, Gaillard JM (1995) Relationship between erythrocyte magnesium, plasma electrolytes and cortisol and intensity of symptoms in major depressed patients. J Affect Disord 34:201–209

98. Linder J, Fyro B, Pettersson U, Werner S (1989) Acute antidepressant effect of lithium is associated with fluctuation of calcium and magnesium in plasma. A double-blind study on the antidepressant effect of lithium and clomipramine. Acta Psychiatr Scand 80:27–36

99. Eby GA, Eby KL (2006) Rapid recovery from major depression using magnesium treatment. Med Hypotheses 67:362–370

100. Barragan-Rodriguez L, Rodriguez-Moran M, Guerrero-Romero F (2008) Efficacy and safety of oral magnesium supplementation in the treatment of depression in the elderly with type 2 diabetes: a randomized, equivalent trial. Magnes Res 21:218–223

101. Pavlinac D, Langer R, Lenhard L, Deftos L (1979) Magnesium in affective disorders. Biol Psychiatry 14:657–661

102. Chouinard G, Beauclair L, Geiser R, Etienne P (1990) A pilot study of magnesium aspartate hydrochloride (Magnesiocard) as a mood stabilizer for rapid cycling bipolar affective disorder patients. Prog Neuropsychopharmacol Biol Psychiatry 14:171–180

103. Cox IM, Campbell MJ, Dowson D (1991) Red blood cell magnesium and chronic fatigue syndrome. Lancet 337:757–760

104. Heiden A, Frey R, Presslich O, Blasbichler T, Smetana R, Kasper S (1999) Treatment of severe mania with intravenous magnesium sulphate as a suppplementary therapy. Psychiatry Res 89:239–246

105. Singewald N, Sinner C, Hetzenauer A, Sartori SB, Murck H (2004) Magnesium-deficient diet alters depression- and anxiety-related behavior in mice – influence of desipramine and Hypericum perforatum extract. Neuropharmacology 47:1189–1197

106. Kantak KM (1988) Magnesium deficiency alters aggressive behavior and catecholamine function. Behav Neurosci 102:304–311

107. Henrotte JG, Franck G, Santarromana M, Frances H, Mouton D, Motta R (1997) Mice selected for low and high blood magnesium levels: a new model for stress studies. Physiol Behav 61:653–658

108. Decollogne S, Tomas A, Lecerf C, Adamowicz E, Seman M (1997) NMDA receptor complex blockade by oral administration of magnesium: comparison with MK-801. Pharmacol Biochem Behav 58:261–268

109. Poleszak E, Szewczyk B, Kedzierska E, Wlaz P, Pilc A, Nowak G (2004) Antidepressant- and anxiolytic-like activity of magnesium in mice. Pharmacol Biochem Behav 78:7–12

110. Poleszak E, Wlaz P, Kedzierska E, Radziwon-Zaleska M, Pilc A, Fidecka S, Nowak G (2005) Effects of acute and chronic treatment with magnesium in the forced swim test in rats. Pharmacol Rep 57:654–658

111. Poleszak E, Wlaz P, Kedzierska E, Nieoczym D, Wrobel A, Fidecka S, Pilc A, Nowak G (2007) NMDA/glutamate mechanism of antidepressant-like action of magnesium in forced swim test in mice. Pharmacol Biochem Behav 88:158–164

112. Cardoso CC, Lobato KR, Binfare RW, Ferreira PK, Rosa AO, Santos AR, Rodrigues AL (2009) Evidence for the involvement of the monoaminergic system in the antidepressant-like effect of magnesium. Prog Neuropsychopharmacol Biol Psychiatry 33:235–242

113. Poleszak E, Wlaz P, Szewczyk B, Kedzierska E, Wyska E, Librowski T, Szymura-Oleksiak J, Fidecka S, Pilc A, Nowak G (2005) Enhancement of antidepressant-like activity by joint administration of imipramine and magnesium in the forced swim test: behavioral and pharmacokinetic studies in mice. Pharmacol Biochem Behav 81:524–529

114. Poleszak E (2008) Benzodiazepine/GABA(A) receptors are involved in magnesium-induced anxiolytic-like behavior in mice. Pharmacol Rep 60:483–489

115. Nowak L, Bregestovski P, Ascher P, Herbet A, Prochiantz A (1984) Magnesium gates glutamate-activated channels in mouse central neurones. Nature 307:462–465

116. Poleszak E (2007) Modulation of antidepressant-like activity of magnesium by serotonergic system. J Neural Transm 114:1129–1134

117. DeVinney R, Wang HH (1995) Mg 2+ enhances high affinity [3H]8-hydroxy-2-(di-n-pro-pylamino) tetralin binding and guanine nucleotide modulation of serotonin -1a receptors. J Recept Signal Transduct Res 15:757–771

Positive Allosteric Modulation of AMPA Receptors: A Novel Potential Antidepressant Therapy

Eric S. Nisenbaum and Jeffrey M. Witkin

Abstract This review focuses on positive allosteric modulation of α-amino-3-hydroxy-5-methyl-4-isoxazolepropionic acid (AMPA) receptors as a novel mechanism to impact mood in the therapy of major depressive disorder (MDD). Several classes of positive allosteric modulators (AMPA receptor potentiators or AMPAkines) have been discovered. Structural and functional studies have demonstrated that different types of positive allosteric modulators have distinct but overlapping binding domains with the extracellular portion of the receptor. Such differences in turn confer unique biophysical mechanisms of allosteric modulation of AMPA receptor function. More recent findings indicate that auxiliary proteins associated with AMPA receptors can further influence the manner in which potentiators ultimately modulate channel function. Support for the hypothesis that AMPA receptor potentiation may be beneficial in MDD comes from both preclinical and clinical findings. Multiple standards of care used to treat MDD have been shown to enhance surface expression of AMPA receptor subunits, increase AMPA subunit phosphorylation, and/or amplify AMPA channel conductance. In addition, AMPA receptor potentiators can enhance brain levels of brain-derived neurotrophic factor, which has been implicated in neuroplasticity and linked to improvements in depression scores after antidepressant treatment. Indeed, neurogenesis has been demonstrated after the application of AMPA receptor potentiators both in vitro and in vivo. Finally, positive allosteric modulators of AMPA receptors are active in a number of rodent models predictive of antidepressant efficacy and where tested, these effects have been prevented by blockade of AMPA receptors. Taken together, biochemical, behavioral, and clinical data suggest that AMPA receptor potentiators may provide a novel therapeutic approach to MDD.

E.S. Nisenbaum (✉)
Neuroscience Discovery Research, Lilly Research Laboratories, Eli Lilly and Company, Lilly Corporate Center, Indianapolis, IN, USA
e-mail: esn@lilly.com

P. Skolnick (ed.), *Glutamate-based Therapies for Psychiatric Disorders*,
Milestones in Drug Therapy, DOI 10.1007/978-3-0346-0241-9_3,
© Springer Basel AG 2010

Abbreviations

AMPA	α-amino-3-hydroxy-5-methyl-4-isoxazolepropionic acid
BDNF	Brain-derived neurotrophic factor
cAMP	Adenosine 3′,5′-cyclic monophosphate
CNQX	6-cyano-7-nitroquinoxaline-2,3-dione
CX516	Piperidine, 1-(6-quinoxalinylcarbonyl)- (9CI); Ampalex; BDP 12
CX691	Piperidine, 1-(2,1,3-benzoxadiazol-5-ylcarbonyl)- (9CI), Farampator
DARPP-32	Dopamine- and cAMP-regulated phosphoprotein of M_r 32,000
DNQX	6,7-dinitroquinoxaline-2,3-dione
GPCR	G protein-coupled receptor
GYKI	53655: 7H-1,3-Dioxolo[4,5-h][2,3]benzodiazepine-7-carboxamide, 5-(4-aminophenyl)-8,9-dihydro-N,8-dimethyl-, monohydrochloride (9CI)
LY392098	N-2-(4-(3-Thienyl)phenyl)propyl 2-propanesulfonamide
LY404187	N-2-(4-(4-Cyanophenyl)phenyl)propyl 2-propanesulfonamide
LY451646	N-[2-(4′-cyano[1,1′-biphenyl]-4-yl)propyl]- (9CI)
LY503430	4′-[(1R)-1-fluoro-1-methyl-2-[[(1-methylethyl)sulfonyl]amino]ethyl]-N-methyl- (9CI)
MGS0039	2-amino-3-[(3,4-dichlorophenyl)methoxy]-6-fluoro-, (1R,2R,3R,5R,6R)- (9CI)
NMDA	N-methyl-D-aspartate

1   Overview

Mood disorders constitute one of the world's major psychiatric maladies. Mood disorders are diagnostically defined by depressed mood, marked reduction in interest and pleasure in life, changes in weight and sleep, a diminished ability to think and make decisions, and suicidal ideation. Formal classification as a mood disorder (DSM-IV of the American Psychiatric Association) requires that at least some of these symptoms are not transient or due to external factors. A number of disorders are subsumed under the category of mood disorders such as bipolar disorders, major depression, masked depression, dysthymic disorder, psychotic depression, atypical depression, and cyclothymic disorder (cf., [1]). The current review focuses on AMPA receptor potentiation as a novel mechanism to impact the symptoms of major depressive disorder (MDD). As there remains significant unmet need in the medicinal therapy for MDD (see [2, 3]), AMPA receptor potentiation could provide an alternative approach to address this considerable gap. The current medicines used to treat depression increase the concentration of biogenic amines (norepinephrine, serotonin, and/or dopamine). This common biochemical action of structurally and pharmacologically diverse antidepressants is one of the foundations of the biogenic amine theory of depression [4, 5]. Although an understanding of the machinery involved in antidepressant efficacy has been evolving, the initiation of antidepressant effects by monoamine-based therapies is thought to occur through the biochemical cascade operating through cAMP signaling (Fig. 1). A neurotoxic hypothesis posits that MDD results from neural insult arising from biochemical stressors (such as glucocorticoids) and that antidepressants engender neurotrophic factors such as brain-derived neurotrophic factor (BDNF) that provide cellular protection, growth, and resilience [7]. By increasing the synaptic availability of biogenic amines, antidepressants might restore the neurochemical milieu with an environment more conducive to creating normal affective tone and adaptability (c.f., [1, 4, 5, 7]). The pathways through which antidepressants impact BDNF have been described along with some inconsistencies in the BDNF model (see [6] for a summary; Fig. 1). Of particular importance, positive allosteric modulation of AMPA receptors is one pathway leading to BDNF production.

It could be postulated that one reason for the relative low percentages of response and remission rates produced by current antidepressants is that nonbiogenic amine-based mechanisms need to be fully engaged. Indeed, it is compelling in this regard that the NMDA receptor antagonist, ketamine, is effective in patients' refractory to conventional antidepressants (see [8]). A role for glutamate neurotransmission in the control of mood has been described [9] and specific modulators of glutamate receptors have been discussed with respect to their potential roles in mood, including NMDA receptors (see [8]) and metabotropic glutamate (mGlu) receptors [10]. An alternative strategy that has been proposed to impact MDD is that of AMPA receptor potentiation that we present here (Fig. 1).

2 AMPA Receptors

AMPA receptors belong to a larger family of ionotropic glutamate receptors which also include N-methyl-D-aspartic acid, NMDA, and kainic acid receptors and can be distinguished on the basis of their molecular and pharmacological properties [11–13]. AMPA receptors are expressed ubiquitously throughout the central nervous system and mediate the majority of rapid excitatory neurotransmission. In addition to their role in postsynaptic depolarization and neuronal firing, AMPA receptors are intimately involved in a variety of other cellular responses, including the recruitment of voltage-gated ion channels and NMDA receptor activity and the development and expression of long-term synaptic plasticity [14], as well as the induction of neurotrophic factors such as BDNF [15].

The AMPA receptor family includes four different genes, termed GLU_{A1-4} (alternatively, GluR1–4 or GluRA–D) that each encode subunits containing a large extracellular NH_2-terminus domain and four hydrophobic domains labeled TM1–TM4 [13]. Evidence indicates that functional AMPA receptors are tetramers that can be generated by the assembly of one or more subunits, yielding either homomeric or heteromeric configurations, respectively [12]. Further diversity

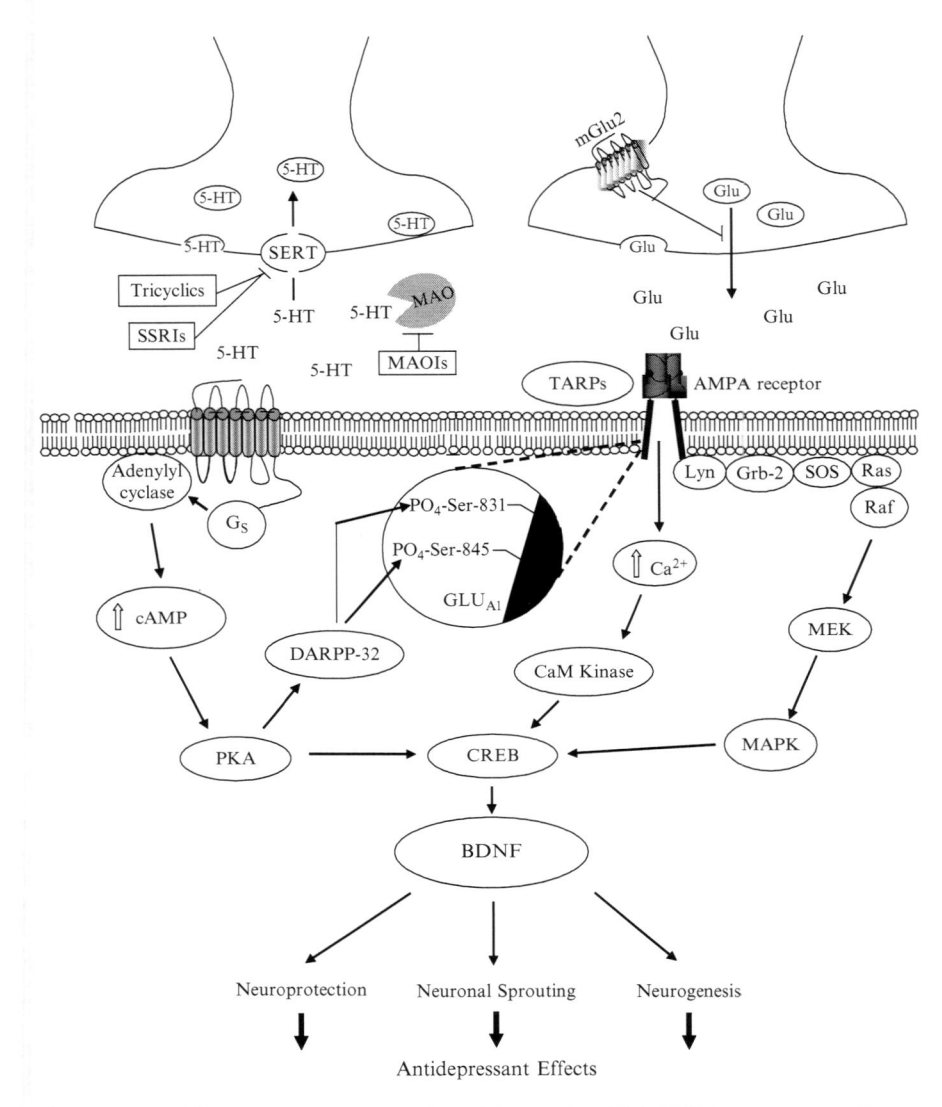

Fig. 1 A simplified schematic representation of the way in which AMPA receptors may impact depression and the action of antidepressant medications. Traditional antidepressants increase synaptic levels of monoamines such as serotonin or norepinephrine (not shown) either by blocking their enzymatic degradation by monoamine oxidase or by blocking their reuptake through the serotonin transporter or norepinephrine transporter (not shown). This leads to activation of PKA pathways which leads to BDNF induction. Monoamine-increasing antidepressants also may induce BDNF through Ca^{2+}/CaM kinase pathways (e.g., through activation of G_q-coupled receptors; not shown). Blockade of 5-HT uptake by the SSRI antidepressant fluoxetine has been shown to phosphorylate the AMPA receptor subunit GLU_{A1} through the DARPP-32 phosphoprotein pathway. The specific serine phosphorylation of GLU_{A1} in turn amplifies the signaling of AMPA receptors. AMPA receptor activity may result in BDNF induction through both Ca^{2+}-dependent and -independent (MAPK) pathways (see text). Postsynaptic AMPA receptors are activated by

among AMPA receptors results from alternative splicing in an extracellular region between TM3 and TM4 in GLU_{A1-4} [16]. This region can contain one of two different exons, referred to as flip (i) and flop (o), which encode a 38 amino acid sequence that differs between the two isoforms by 7–10 residues depending on the subunit. Functionally, flip and flop isoforms of homomeric GLU_{A2-4} receptors display markedly different kinetics of desensitization in the continued presence of glutamate with the flip isoform desensitizing two- to fivefold more slowly than the flop variant for these subunits [17].

On the basis of recent structure–function studies, a model of the mechanisms by which AMPA receptors gate synaptic current has emerged. Evidence indicates that the tetrameric receptor is formed by dimerization of two ligand binding cores (LBC) of adjacent subunits that in turn dimerize with the dimer of the other two LBCs [18–20]. Activation of AMPA receptors is initiated by binding of glutamate to each of the four LBCs, leading to the closure of the upper and lower domains of each LBC, which is predicted to direct a conformational change in the ion channel gate, permitting ion flux through the channel pore [20]. Termination of current flow can occur via two mechanisms. Deactivation reflects closing of the channel upon removal of glutamate and subsequent release of agonist from the LBC. Desensitization occurs in the continued presence of glutamate and reflects a destabilization and subsequent rearrangement of the intradimer interface that is predicted to uncouple the closing of the LBCs from the channel gate, permitting closure of the ion channel. Interestingly, crystallographic data have revealed that the flip/flop cassette in the S2 region is located at the intradimer interface and is associated with the "hinge" of the clamshell region of the LBC [20]. Differences in the desensitization rates of flip and flop receptors have been shown to depend on the identity of only three amino acid residues in the flip–flop region (Thr765, Pro766, and Ser775 in flip and Asn765, Ala766, and Asn775 in flop) and these three residues are postulated to confer their kinetic differences by directly and/or indirectly influencing the stability of the dimer interface between adjacent subunits [17].

AMPA receptor expression and function is also highly regulated by scaffolding proteins to which they are associated. The insertion and removal of AMPA receptors from synapses, which in turn controls excitatory tone and synaptic plasticity, are under complex regulatory control that is only beginning to be understood (see [21, 22] for a review). Among the many proteins that can associate with AMPA receptors, the transmembrane AMPA receptor regulatory proteins (TARPs) are of particular interest given their role in receptor trafficking, expression, and synaptic function. Previous studies have identified six mammalian TARPs, γ-2 (or stargazin), γ-3, γ-4, γ-5, γ-7, and γ-8, that either regulate AMPA receptor expression and/ or enhance AMPA receptor function [22–24]. These proteins display a striking

---

Fig.1 (continued) glutamate, the release of which is regulated by presynaptic metabotropic glutamate receptors, such as $mGlu_2$, which inhibits glutamate release. The details within these major pathways have not been elaborated in this rendition. Further, other pathways are likely. Acitvation of PKA for example could induce CREB through MAPK-mediated mechanisms. Reprinted from [6] with permission from Elsevier

differential expression in the brain [24]. For example, γ-8 is expressed preferentially in the hippocampus [24] and plays a critical role in AMPA receptor expression as evidenced by the marked reduction ($>60\%$ relative to control levels) in the number of synaptic AMPA receptors in the CA1 region of the hippocampus of γ-8 knockout mice [25]. The γ-8 TARP also dramatically influences the kinetics of AMPA receptor gating by attenuating the rate of channel deactivation and desensitization, which in turn controls the time course of synaptic transmission [25]. Interestingly, chronic administration of antidepressants increases the synaptic levels of AMPA receptors in the hippocampus, concomitant with an increase in BDNF mRNA expression [26, 27]. Although the link between hippocampal BDNF, AMPA receptors, and MDD is only beginning to be elucidated, evaluation of the action of antidepressant drugs on BDNF expression, synaptic transmission/plasticity, and/or behavioral function in γ-8-deficient mice could provide important insights into the role of this TARP in the therapeutic benefit of antidepressant drugs.

3 AMPA Receptor Potentiators

The initial discovery of AMPA receptor potentiator molecules came years after the development of a pyrrolidinone nootropic agent, aniracetam when it was found to selectively enhance AMPA receptor-mediated currents in the presence of agonist, but was without effect on its own [28]. Several structurally distinct molecules were subsequently discovered that function as positive allosteric modulators of AMPA receptors. AMPA receptor potentiators include benzothiazides such as cyclothiazide, benzylpiperidines (e.g., CX516, CX546), and biarylpropylsulfonamides as exemplified by LY392098 and LY503430 [29–31].

AMPA receptor potentiators can have distinct effects on the biophysical processes of deactivation and desensitization. Aniracetam preferentially attenuates the deactivation of AMPA receptor currents with little effect on receptor desensitization, whereas cyclothiazide almost exclusively slows the transition into the desensitized state of the receptor [29, 32]. These distinct biophysical mechanisms of action have different functional consequences on synaptic transmission. For example, CX516, which primarily slows deactivation, markedly enhances synaptic responses in hippocampal neurons to single stimuli; in contrast, cyclothiazide principally increases synaptic inputs in response to high frequency stimulation [33, 34]. Compounds that affect both deactivation and desensitization are predicted to produce the most robust augmentation of synaptic transmission.

AMPA receptor potentiators also can preferentially enhance current flux through either the flip or flop isoforms of homomeric AMPA receptors. Cyclothiazide displays approximately a 50-fold greater potency for enhancing currents through GLU_{A1} flip receptors compared with GLU_{A1} flop receptors [29, 32]. A similar preference for flip splice variants has been described for LY503430 [35]. In contrast, aniracetam, and the sulfonylamino compound, PEPA, are considerably more potent at flop variants of GLU_{A1-4} homomeric receptors [29, 36]. At first glance, these data suggest that compounds with splice variant selectivity may be capable of targeting AMPA receptors subtypes that are preferentially expressed in particular brain regions, leading to regional enhancement of glutamatergic synaptic transmission. However, more recent data indicates the isoform selectivity of allosteric modulators that can be influenced by the presence of TARPs. For example, coexpression of γ-2 with Glu_{A1} flip or flop receptors changes their pharmacology, making both isoforms sensitive to either cyclothiazide or PEPA [37]. In addition, coexpression of γ-2 with Glu_{A1} uncovers an effect of cyclothiazide on deactivation – not seen in the absence of the TARP – on both flip and flop isoforms of the receptor. TARPs also can influence the effects of potentiators on agonist binding to Glu_A receptors [37]. Recent studies have shown that the maximal binding of [^{3}H] fluorowillardiine to Glu_{A2} receptors in the presence of CX546 is significantly increased by coexpression with either γ-3 or γ-8 [38]. Consistent with this finding, CX546 enhances agonist binding differentially in discrete brain regions such that the largest increases are found in hippocampus, where Glu_{A2}, γ-3, and γ-8 are highly expressed [38]. Collectively, these data indicate that TARPs can dramatically influence the biophysical mechanisms of action, isoform selectivity, as well as the agonist affinity of AMPA receptor potentiators. In light of the heterogeneous distribution of TARPs in brain, these data suggest that identification of TARP-dependent AMPA receptor potentiators may provide a novel mechanism for preferentially modulating specific synapses within the central nervous system.

4 AMPA Receptor Potentiators Modulate Antidepressant-Related Biochemistry and Structural Biology

A prevailing hypothesis underlying the mechanisms of depression and antidepressant action involves the regulation and expression of growth and neurotrophic factors [6, 7, 39] as illustrated in Fig. 1. BDNF has been the focus of many studies, although other factors have been explored, including glial-derived neurotrophic factor (GDNF) and vascular–endothelial growth factor (VEGF) (c.f., [2]). The biology of BDNF is complex and involves several isoforms, which can recruit multiple signaling pathways [40–43]. The primary biochemical processes regulated by BDNF are initiated through the binding of BDNF to the full-length catalytic TrkB receptor. The TrkB pathways function to regulate cellular survival and differentiation and it is their role in neurogenesis that BDNF is thought to confer antidepressant activity [44]. BDNF can have both neurotrophic and neuroprotective effects (e.g., [45–47]), and is a likely promoter of neurogenesis [6]. It is hypothesized that the neurotrophic and neurogenic actions of BDNF may produce antidepressant effects by reversing the morphological deficits that might be associated with depression [6, 7]. Indeed, neurogenesis has been experimentally linked to the antidepressant-like effects of drugs in rodent models [48].

A role for BDNF in MDD comes from convergent data sources. Clinical studies have demonstrated alterations in BDNF levels in patients and changes in BDNF levels produced by antidepressants concomitant with improvements in mood. Shimizu et al. [49], showed that serum BDNF levels were significantly lower in antidepressant-naive patients with MDD than in either normal control subjects or antidepressant-treated patients with MDD. In addition, a negative association between serum BDNF and scores on the Hamilton depression scale was observed in this study.

Further support for the BDNF hypothesis of MDD comes from preclinical studies that have shown that administration of antidepressants can alter BDNF levels in brain and that administration of BDNF can produce antidepressant-like changes in biochemical and behavioral endpoints, some of which are eliminated in BDNF transgenic animals. The BDNF hypothesis of antidepressant action arose from the initial observation that a variety of antidepressant drugs were able to increase the mRNA encoding BDNF and its receptor, TrkB, in rat hippocampus following chronic treatment [50]. In addition to conventional antidepressants, the expression of BNDF mRNA and/or protein was also increased following treatment with a phosphodiesterase inhibitor [51, 52], electroconvulsive shock [50], or chronic exercise [53, 54], all of which can be efficacious in the treatment of depression. Based on these convergent data, it has been hypothesized that BDNF induction may be a common mechanism by which these various treatments produce their antidepressant efficacy. Additional support for this hypothesis comes from the finding that direct infusion of recombinant BDNF into the central nervous system can produce antidepressant-like effects in rats in both the forced swim and learned helplessness models [6, 7, 55]. Antidepressants were reported to have no behavioral effect in the forced swim test in heterozygous BDNF knockout mice, or in mutant mice expressing a dominant negative isoform TrkB, whereas the antidepressants were effective in wild-type mice [56].

As reported with conventional antidepressants, AMPA receptor potentiators also increase BDNF expression (Table 1 for summary). Early studies showed that treatments with the positive allosteric modulator, CX614 markedly and reversibly increased BDNF mRNA and protein levels in cultured rat hippocampal slices [15]. Subsequent studies demonstrated that chronic treatment (7 days) with LY451646 increased the level of both BDNF and TrkB mRNA expression in the dentate gyrus, CA3 and CA4 of the hippocampus [58]. In addition, LY451646 has been shown to increase progenitor cell proliferation in adult rat hippocampus [59]; rats treated for 21 days with LY451646 displayed a 45% increase in the number of cells labeled with bromodeoxyuridine (BrdU), a marker of mitosis. However, cellular growth and differentiation were not assessed in this study.

A model for the induction of BDNF expression is provided in Fig. 1. Conventional antidepressants induce BDNF expression by activating the transcription factor Ca^{2+}/cAMP response element binding protein (CREB) through protein kinase A (PKA) and Ca^{2+}/calmodulin-dependent protein kinase (CaM kinase) signal transduction pathways [6, 67–69]. AMPA receptor potentiators can also induce BDNF by activating CaM kinase signal transduction pathways. In addition, AMPA receptor potentiators can engender BDNF production through a Ca^{2+}-independent mitogen-activated protein kinase (MAPK) pathway. Lyn, a member

Table 1 Effects of AMPA receptor potentiators on antidepressant-related biochemistry and behavior

Compound	Method – species	Findings	Reference
CX546, CX614	Entorhinal/hippocampal slices – rat	Increased BDNF and trkB mRNA	[15]
CX546	Hippocampus – rat	Increased BDNF mRNA	[15]
LY392098	Cortical neurons – rat	Increased BDNF	[57]
LY451646	Hippocampus – rat	Increased BDNF	[58]
LY451646	Hippocampus – rat	Increased cell proliferation	[59]
Org 26576	Hippocampus – rat	Increased cell proliferation and survival	[60]
LY392098	Forced-swim – mouse	Decreased immobility	[61, 62]
LY392098	Forced-swim – rat	Decreased immobility	[62]
LY392098	Tail suspension – mouse	Decreased immobility	[62]
LY392098	Forced-swim – mouse	Synergy with conventional antidepressants	[61]
LY392098	Chronic-mild stress – mouse	Effects on multiple endpoints (see Table 3)	[63]
LY404187, LY451646	Tail suspension – Mouse	Decreased immobility	[64]
LY404187, LY451646	Forced swim – mouse	Decreased immobility	[64]
CX516, CX691, CX731	Submissive behavior – rat	Decreased submissive behavior	[65]
CX614	Forced swim – mouse and rat	Decreased immobility	[66]

of the Src family of protein tyrosine kinases, is physically associated with AMPA receptor subunits, and can be activated by AMPA receptor activation. Lyn activation is independent of Ca^{2+} flux or neuronal firing, and leads to MAPK activation and an increase in BDNF mRNA [70]. The AMPA receptor potentiator LY392098 has been shown to increase the expression of BDNF mRNA in rat cortical neurons in both a Ca^{2+}-dependent and a Ca^{2+}-independent, nimodipine-insensitive manner, which appears to be mediated by a MAPK pathway [57].

5 AMPA Receptors Are Modulated in Depression and by Antidepressant Treatment

Further support for the notion that AMPA receptor potentiators may provide antidepressant actions comes from the limited findings of AMPA receptor alterations in mood disorder patients, as well as AMPA receptor modification by antidepressant drugs. Increases in AMPA receptor binding have been reported in the striatum of suicide victims [71, 72]. However, in both human studies and in animals exposed to stressors, both increases and decreases in AMPA receptors and AMPA receptor-related transduction pathways have been observed (see [6, 73–81]). With conventional antidepressants, the existent data is generally consistent with the findings that subchronic treatment with antidepressants upregulates the function

Table 2 Effects of antidepressants on AMPA receptor function

Compound	Method – species	Findings	Reference
Desipramine Paroxetine	Hippocampus – rat	GLU_{A1} and $GLU_{A2/3}$ increased	[26]
Imipramine	Hippocampal synaptosomes – rat	GLU_{A1} increased	[82]
Electroconvulsive therapy	Hippocampus ? – rat	GLU_{A1} message increased	[83]
Fluoxetine	Hippocampus, striatum, and cortex – mouse	Phosphorylation of GLU_{A1} increased	[84]
Tianeptine	Frontal cortex and CA3 of hippocampus – mice	Phosphorylation of GLU_{A1} increased Effect absent in Phoshomutant mice	[85]
Imipramine Lamotragine Riluzole	Cultured hippocampal neurons	GLU_{A1} and GLU_{A2} increased Peak AMPA currents increased	[86]
Imipramine Lamotragine Riluzole	Hippocampus – mouse	Phosphorylation of GLU_{A1} increased	[86]

of AMPA receptors (Table 2). For example, fluoxetine phosphorylates Ser^{845} in GLU_{A1} through PKA which can result in augmentation of peak current amplitude in homomeric GLU_{A1} receptor channels [87]. Likewise, AMPA receptor currents can also be potentiated by CaM kinase II or protein kinase C, which increase the phosphorylation of Ser^{831} [88]. The ratio of phosphorylation states of Ser^{831} and Ser^{845} has been shown to regulate synaptic plasticity [89]; alteration of this ratio, as occurs upon chronic dosing with fluoxetine [84], might therefore represent a mechanism driving antidepressant efficacy.

In contrast to conventional antidepressants, mood stabilizers generally lead to decreased AMPA receptor function [6]. A recent report [82, 86, 90] studying the effects of three anticonvulsants used in the treatment of bipolar disorder on AMPA receptor expression and function supports this perspective. The anticonvulsants that have marked antidepressant efficacy, lamotrigine and riluzole, had positive effects on the amplification of AMPA receptor function, including increases in surface expression of GLU_{A1} and GLU_{A2}, peak AMPA current, and phosphorylation of GLU_{A1} at Ser845 after repeat drug exposure. In contrast, with valproate, which is primarily used as an antimania treatment, generally the opposite effects were noted.

Imaging in humans undergoing antidepressant treatments have substantiated the enhancement in glutamate utilization that occurs under these conditions. For example, when dosed with riluzole, bipolar patients exhibit increases in glutamate/glutamine ratios, a measure of the turnover or utilization of glutamate [91]. From such observations, it can also be deduced that downstream targets for glutamate will be amplified. Data are accumulating to suggest that the ratio of AMPA to NMDA function is enhanced at least in the initiation of antidepressant activity [92].

6 AMPA Receptor Potentiators Engender Antidepressant-Like Behavioral Effects

Animal models are often used to predict efficacy of novel mechanisms in the treatment of MDD. Several AMPA receptor potentiators have demonstrated efficacy comparable to that of conventional biogenic amine antidepressants under a host of conditions (Table 1). Results include positive findings in both acute assays such as the forced-swim test and in studies requiring subchronic dosing. For example, antidepressant-like efficacy has been observed for the AMPA receptor potentiators, piracetam, aniracetam, CX516 (BDP12), CX691, and CX731 in a rat dominant/submissive assay. Similar to the effects of fluoxetine, these AMPA receptor potentiators decreased the amount of submissive behavior of rats when given subchronically [65]. In addition, the onset of action of the antidepressant-like effects of the AMPA receptor potentiators was faster under these conditions than that for fluoxetine. In another model, utilizing subchronic dosing, Farley et al. [63] showed that LY392098 effectively prevented several behaviors arising from chronic mild stress in mice. Thus, AMPA receptor potentiators of two structural classes have demonstrated antidepressant-like efficacy in mice and rats, when tested under either acute or subchronic dosing regimens.

Since many patients do show remission with antidepressants, add-on therapies are one method used for potential augmentation. Only a few agents have been approved for use as add-on therapies for MDD, the latest being the antipsychotic aripiprazole [2]. Interestingly, AMPA receptor potentiators produce synergistic effects when combined with clinically effective antidepressants. Using the mouse forced swim test, ineffective doses of the AMPA receptor potentiator LY392098 significantly augmented the effects of other antidepressant compounds with distinct mechanisms of action, including SSRIs (fluoxetine and citalopram), a norepinephrine uptake inhibitor (nisoxetine), a mixed norepinephrine/serotonin uptake blocker (duloxetine), a tricyclic antidepressant (imipramine), and a phosphodiesterase 4 inhibitor (rolipram). Likewise, ineffective doses of the traditional antidepressants potentiated the antidepressant-like effects of LY392098 [61]. These observations are consistent with the idea that AMPA receptor potentiators produce their antidepressant-like effects through a mechanism that, although distinct, ultimately converges upon a common final pathway (Fig. 1).

If positive allosteric modulators of AMPA receptors are a novel mechanism-based antidepressant, then their behavioral and biochemical actions should have points of disjunction from that of the biogenic amine-based antidepressants. Indeed, the AMPA receptor potentiator LY392098 does not directly impact monoamine levels in cortex of rats as observed from in vivo microdialysis experiments [3]. In addition, the mechanism of action of LY392098 in the mouse forced-swim test has been demonstrated to be due to AMPA receptor potentiation. The decrease in immobility in the forced-swim test in mice produced by LY392098 was prevented by prior administration of an AMPA receptor antagonist. In contrast, a comparable antidepressant-like effect induced by acute dosing with the tricyclic antidepressant, imipramine was not significantly prevented by AMPA receptor blockade [62].

A recent study provides additional support for the AMPA receptor potentiation hypothesis of antidepressant action. Mice exposed to chronic mild stress exhibit a number of behavioral changes. Subchronically-dosed fluoxetine prevented all of these changes in behavior, whereas LY392098 eliminated only a subset of those behaviors (Table 3). Notably, the AMPA receptor antagonist, NBQX, blocked the antidepressant-like effects of fluoxetine only for those behaviors that were positively impacted by LY392098, but not when given alone. The observation that subchronic dosing with fluoxetine produces in vivo effects that are attenuated by an AMPA receptor antagonist provides the first evidence demonstrating the AMPA receptor dependence of the antidepressant-like behavioral effects of a biogenic-amine-based antidepressant using AMPA antagonists. The prior report by Li et al. [62] utilized acute dosing with imipramine in the forced swim test, where AMPA receptor blockade was ineffective in preventing its antidepressant-like effects. In the work by Farley et al. [63] subchronic dosing was studied under a chronic mild stress paradigm. It is speculated that the ability of NBQX to block behavioral effects of fluoxetine in the later study are due to the differential changes in AMPA receptor phosphorylation that accompanies subchronic vs. acute dosing [84].

Other antidepressants and putative antidepressants also might confer antidepressant-like effects through a mechanism involving potentiation of AMPA receptors. Table 4 summarizes data in multiple animal models, where antidepressant-like biochemical and behavioral effects of known and putative antidepressants have been shown to be prevented by AMPA receptor antagonists. Note that there are a few cases where blockade by an AMPA receptor antagonist was not achieved.

7 Conclusions

An accumulation body of evidence is emerging suggesting that AMPA receptor potentiators might provide a novel therapeutic approach to MDD. Positive

Table 3 Effects of the AMPA receptor potentiator LY392098 and fluoxetine on behaviors after chronic mild stress in mice

Sign or effect	Fluoxetine	LY392098	NBQX block of fluoxetine
Weight loss	+	+	+
Fur deterioration	+	+	+
Immobility in TST	+	+	+
Sucrose preference	+	−	−
Marble-burying	+	−	−

Data are summarized from [63]. Fluoxetine was given at 20 mg/kg/day and LY392098 at a dose of 5 mg/kg/day
TST tail-suspension test

Table 4 AMPA receptor dependence of the antidepressant-like effects of antidepressants and putative antidepressants

Compound (mechanism)	Method – species	Findings	Reference
MGS0039 (mGlu2/3 antagonist)	Microdialysis – medial prefrontal cortex – rat	Increased serotonin efflux blocked by NBQX	[93]
MGS0039 (mGlu2/3 antagonist)	Tail-suspension test – mouse	Antidepressant-like effect prevented by NBQX	[93]
LY392098 (AMPA potentiator)	Forced-swim test – mouse	Antidepressant-like effect prevented by GYKI3655	[62]
LY392098 (AMPA potentiator)	Various behaviors	See Table 3	[63]
LY392098 (AMPA potentiator)	Primary neuron culture – rat	Increased BDNF mRNA blocked by NBQX	[57]
Imipramine – acute (Monoamine uptake blocker)	Forced-swim test – mouse	Acute antidepressant-like effects not prevented by NBQX	[62]
Fluoxetine – subchronic (Monoamine uptake blocker)	Various behaviors	See Table 3	[63]
CX614 (AMPA potentiator)	Forced-swim test – mouse	Antidepressant-like effect prevented by NBQX	[66]
Ketamine (Uncompetitive NMDA antagonist)	Forced-swim test – mouse	Antidepressant-like effect prevented by NBQX	[94]
Zinc (Uncompetitive NMDA antagonist)	Forced-swim test – mouse	Antidepressant-like effect prevented by NBQX	[66]
CGP37849 (Competitive NMDA antagonist)	Forced-swim test – mouse	Antidepressant-like effect not prevented by NBQX	[95]
Lithium	Forced-swim test – mouse	Antidepressant-like effect prevented by NBQX	[96]

allosteric modulators of AMPA receptors increase brain levels of BDNF which has been implicated in the etiology of MDD, as well as the efficacy of conventional antidepressant therapies. AMPA receptor potentiators are effective in acute and chronic models of antidepressant activity. In addition, mutually synergistic effects of AMPA receptor potentiators and conventional antidepressants have been demonstrated in animal models. Although direct tests of the AMPA potentiation hypothesis of MDD have yet to be reported, the convergence of data strongly suggest that continued experimental attention to the role of AMPA receptors in mood disorders is warranted. It is hoped that with such experimental scrutiny, the needed improvements in the medicines will be enabled for the devastating disorders of mood.

Acknowledgments We are also grateful to our many colleagues for discussion in particular to Drs. Phil Skolnick, Darryle Schoepp, James Monn, Paul Ornstein, Eleni Tzavara, David Bleakman, and David Bredt.

References

1. Dubovsky SL, Buzan R (1999) Mood disorders. In: Hales RE, Yudofsky SC, Talbott JA (eds) Textbook of psychiatry, 3rd edn. American Psychiatric Press, Washington, DC, pp 479–565
2. Witkin JM, Li X (2009) New approaches to the management of major depressive disorder. In: Enna SJ, Williams M (eds) Contemporary aspects of biomedical research: biomedical issues. advances in pharmacology, vol 57. Academic, New York, pp 347–379
3. Skolnick P, Legutko B, Li X, Bymaster FP (2001) Current perspectives on the development of non-biogenic amine-based antidepressants. Pharmacol Res 43:411–443
4. Millan MJ (2004) The role of monamines in the actions of established and "novel" antidepressant agents: a critical review. Eur J Pharmacol 500:371–384
5. Iversen L (2005) The monoamine hypothesis of depression. In: Licinio J, Wong ML (eds) Biology of depression. Wiley-VCH, Weinheim, pp 71–86
6. Alt A, Nisenbaum ES, Bleakman D, Witkin JM (2006) A role for AMPA receptors in mood disorders. Biochem Pharmacol 71:1273–1288
7. Duman RS (2004) The neurochemistry of depressive disorders: preclinical studies. In: Charney DS, Nestler EJ (eds) Neurobiology of mental illness, 2nd edn. Oxford University Press, Oxford, pp 421–439
8. Skolnick P, Popik P, Trullas R (2010) N-Methyl-D-aspartate (NMDA) antagonists for the treatment of Depression. In: Skolnick P (ed) Glutamate-based therapies for psychiatric disorders. Springer, Basel
9. Schoepp DD, Skolnick P (2010) Introduction. In: Skolnick P (ed) Glutamate-based therapies for psychiatric disorders. Springer, Basel
10. Witkin JM (2010) Glutamatergic Modulators for the Treatment of Major Depressive Disorder: Metabotropic Glutamate Receptors. In: Skolnick P (ed) Glutamate-based therapies for psychiatric disorders. Springer, Basel
11. Bleakman D, Lodge D (1998) Neuropharmacology of AMPA and kainate receptors. Neuropharmacology 37:1187–1204
12. Dingledine R, Borges K, Bowie D, Traynelis SF (1999) The glutamate receptor ion channels. Pharmacol Rev 51:7–61
13. Hollmann M, Heinemann S (1994) Cloned glutamate receptors. Annu Rev Neurosci 17:31–108
14. Malenka RC, Nicoll RA (1999) Long-term potentiation – a decade of progress? Science 281:1870–1874
15. Lauterborn JC, Lynch G, Vanderklish P, Arai A, Gall CM (2000) Positive modulation of AMPA receptors increases neutrophin expression by hippocampal and cortical neurons. J Neurosci 20:8–21
16. Sommer B, Keinänen K, Verdoorn TA, Wisden W, Burnashev N, Herb A, Köhler M, Takagi T, Sakmann B, Seeburg PH (1990) Flip and flop: a cell-specific functional switch in glutamate-operated channels of the CNS. Science 249:1580–1585
17. Quirk JC, Siuda ER, Nisenbaum ES (2004) Molecular determinants responsible for differences in desensitization kinetics of AMPA receptor splice variants. J Neurosci 24:11416–11420
18. Armstrong N, Sun Y, Chen GQ, Gouaux E (1998) Structure of a glutamate-receptor ligand-binding core in complex with kainate. Nature 395:913–917
19. Armstrong N, Gouaux E (2000) Mechanisms for activation and antagonism of an AMPA-sensitive glutamate receptor: crystal structures of the GluR2 ligand binding core. Neuron 28:165–181
20. Sun Y, Olson R, Horning M, Armstrong N, Mayer M, Gouaux E (2002) Mechanism of glutamate receptor desensitization. Nature 417:245–253
21. Esteban JA (2003) AMPA receptor trafficking: a road map for synaptic plasticity. Mol Interv 3:375–385

22. Bredt DS, Nicoll RA (2003) AMPA receptor trafficking at excitatory synapses. Neuron 40:361–379
23. Kato AS, Siuda ER, Nisenbaum ES, Bredt DS (2008) AMPA receptor subunit-specific regulation by a distinct family of type II TARPs. Neuron 59:986–996
24. Tomita S, Chen L, Kawasaki Y, Petralia RS, Wenthold RJ, Nicoll RA, Bredt DS (2003) Functional studies and distribution define a family of transmembrane AMPA receptor regulatory proteins. J Cell Biol 161:805–816
25. Milstein A, Zhou W, Karimzadegan S, Bredt D, Nicoll R (2007) TARP subtypes differentially and dose-dependently control synaptic AMPA receptor gating. Neuron 55:905–918
26. Martinez-Turrillas R, Frechilla D, Del Rio J (2002) Chronic antidepressant treatment increases the membrane expression of AMPA receptors in rat hippocampus. Neuropharmacology 43:1230–1237
27. Martínez-Turrillas R, Del Río J, Frechilla D (2005) Sequential changes in BDNF mRNA expression and synaptic levels of AMPA receptor subunits in rat hippocampus after chronic antidepressant treatment. Neuropharmacology 49:1178–1188
28. Ito I, Tanabe S, Kohda A, Sugiyama H (1990) Allosteric potentiation of quisqualate receptors by a nootropic drug aniracetam. J Physiol 424:533–543
29. Partin KM, Fleck MW, Mayer ML (1996) AMPA receptor flip/flop mutants affecting deactivation, desensitization, and modulation by cyclothiazide, aniracetam, and thiocyanate. J Neurosci 16:6634–6647
30. Miu P, Jarvie KR, Radhakrishnan V, Gates MR, Ogden A, Ornstein PL (2001) Novel AMPA receptor potentiators LY392098 and LY404187: effects on recombinant human AMPA receptors in vitro. Neuropharmacology 40:976–983
31. Quirk JC, Nisenbaum ES (2002) LY404187: a novel positive allosteric modulator of AMPA receptors. CNS Drug Rev 8:255–282
32. Partin KM, Patneau DK, Mayer ML (1994) Cyclothiazide differentially modulates desensitization of AMPA receptor splice variants. Mol Pharmacol 46:129–138
33. Arai A, Lynch G (1998) AMPA receptor desensitization modulates synaptic responses induced by repetitive afferent stimulation in hippocampal slices. Brain Res 799:235–242
34. Arai A, Lynch G (1998) The waveform of synaptic transmission at hippocampal synapses is not determined by AMPA receptor desensitization. Brain Res 799:230–234
35. O'Neill MJ, Murray TK, Clay MP, Lindstrom T, Yang CR, Nisenbaum ES (2005) LY503430: pharmacology, pharmacokinetics, and effects in rodent models of Parkinson's disease. CNS Drug Rev 11:77–96
36. Sekiguchi M, Fleck MW, Mayer ML, Takeo J, Chiba Y, Yamashita S et al (1997) A novel allosteric potentiator of AMPA receptors: 4-[2-(phenylsulfonylamino)ethylthio]-2, 6-difluoro-phenoxyacetamide. J Neurosci 17:5760–5771
37. Tomita S, Sekiguchi M, Wada K, Nicoll RA, Bredt DS (2006) Stargazin controls the pharmacology of AMPA receptor potentiators. Proc Natl Acad Sci USA 103:10064–10067
38. Montgomery KE, Kessler M, Arai AC (2009) Modulation of agonist binding to AMPA receptors by 1-(1, 4-benzodioxan-6-ylcarbonyl)piperidine (CX546): differential effects across brain regions and GluA1–4/transmembrane AMPA receptor regulatory protein combinations. J Pharmacol Exp Ther 331:965–974
39. Castrén E, Rantamäki T (2010) Role of brain-derived neurotrophic factor in the aetiology of depression: implications for pharmacological treatment. CNS Drugs 24:1–7
40. Chao MV (2003) Neurotrophins and their receptors: a convergence point for many signalling pathways. Nat Rev Neurosci 4:299–309
41. Schlessinger J, Ullrich A (1992) Growth factor signaling by receptor tyrosine kinases. Neuron 9:383–391
42. Allendoerfer KL, Cabelli RJ, Escandon E, Kaplan DR, Nikolics K, Shatz CJ (1994) Regulation of neurotrophin receptors during the maturation of the mammalian visual system. J Neurosci 14:1795–1811

43. Middlemas DS, Lindberg RA, Hunter T (1991) trkB, a neural receptor protein-tyrosine kinase: evidence for a full-length and two truncated receptors. Mol Cell Biol 11:143–153
44. Wang JW, Dranovsky A, Hen R (2008) The when and where of BDNF and the antidepressant response. Biol Psychiatry 63:640–641
45. Frim DM, Uhler TA, Galpern WR, Beal MF, Breakefield XO, Isacson O (1994) Implanted fibroblasts genetically engineered to produce brain-derived neurotrophic factor prevent 1-methyl-4-phenylpyridinium toxicity to dopaminergic neurons in the rat. Proc Natl Acad Sci USA 91:5104–5108
46. Mamounas LA, Blue ME, Siuciack JA, Altar CA (1995) Brain-derived neurotrophic factor promotes the survival and sprouting of serotonergic axons in rat brain. J Neurosci 15: 7929–7939
47. Mamounas LA, Altar CA, Blue ME, Kaplan DR, Tessarollo L, Lyons WE (2000) BDNF promotes the regenerative sprouting, but not survival, of injured serotonergic axons in the adult rat brain. J Neurosci 20:771–782
48. Santarelli L, Saxe M, Gross C, Surget A, Battaglia F, Dulawa S et al (2003) Requirement of hippocampal neurogenesis for the behavioral effects of antidepressants. Science 301:805–809
49. Shimizu E, Hashimoto K, Okamura N, Koike K, Komatsu N, Kumakiri C et al (2003) Alterations of serum levels of brain-derived neurotrophic factor (BDNF) in depressed patients with or without antidepressants. Biol Psychiatry 54:70–75
50. Nibuya M, Morinobu S, Duman RS (1995) Regulation of BDNF and trkB mRNA in rat brain by chronic electroconvulsive seizure and antidepressant drug treatments. J Neurosci 15:7539–7547
51. Nibuya M, Nestler EJ, Duman RS (1996) Chronic antidepressant administration increases the expression of cAMP response element binding protein (CREB) in rat hippocampus. J Neurosci 16:2365–2372
52. Fujimaki K, Morinobu S, Duman RS (2000) Administration of a cAMP phosphodiesterase 4 inhibitor enhances antidepressant-induction of BDNF mRNA in rat hippocampus. Neuropsychopharmacology 22:42–51
53. Neeper SA, Gómez-Pinilla F, Choi J, Cotman CW (1995) Exercise and brain neurotrophins. Nature 373:109
54. Neeper SA, Gómez-Pinilla F, Choi J, Cotman CW (1996) Physical activity increases mRNA for brain-derived neurotrophic factor and nerve growth factor in rat brain. Brain Res 726:49–56
55. Shirayama Y, Chen AC, Nakagawa S, Russell DS, Duman RS (2002) Brain-derived neurotrophic factor produces antidepressant effects in behavioral models of depression. J Neurosci 22:3251–3261
56. Saarelainen T, Hendolin P, Lucas G, Koponen E, Sairanen M, MacDonald E et al (2003) Activation of the TrkB neurotrophin receptor is induced by antidepressant drugs and is required for antidepressant-induced behavioral effects. J Neurosci 23:349–357
57. Legutko B, Li X, Skolnick P (2001) Regulation of BDNF expression in primary neuron culture by LY392098, a novel AMPA receptor potentiator. Neuropharmacology 40:1019–1027
58. Mackowiak M, O'Neill M, Hicks C, Bleakman D, Skolnick P (2002) An AMPA receptor potentiator modulates hippocampal expression of BDNF: An in vivo study. Neuropharmacology 43:1–10
59. Bai F, Bergeron M, Nelson DL (2003) Chronic AMPA receptor potentiator (LY451646) treatment increases cell proliferation in adult rat hippocampus. Neuropharmacology 44:1013–1021
60. Su XW, Li XY, Banasr M, Koo JW, Shahid M, Henry B, Duman RS (2009) Chronic treatment with AMPA receptor potentiator Org 26576 increases neuronal cell proliferation and survival in adult rodent hippocampus. Psychopharmacol (Berl) 206:215–222
61. Li X, Witkin JM, Need AB, Skolnick P (2003) Enhancement of antidepressant potency by a potentiator of AMPA receptors. Cell Mol Neurobiol 23:419–430

62. Li X, Tizzano JP, Griffey K, Clay M, Lindstrom T, Skolnick P (2001) Antidepressant-like actions of an AMPA receptor potentiator (LY392098). Neuropharmacology 40:1028–1033
63. Farley S, Apazoglou K, Witkin JM, Giros B, Tzavara ET (2010) Antidepressant-like effects of an AMPA receptor potentiator under a chronic mild stress paradigm. Int J Neuropsychopharmacol 11:1–12
64. Bai F, Li X, Clay M, Lindstrom T, Skolnick P (2001) Intra- and interstrain differences in models of "behavioral despair". Pharmacol Biochem Behav 70:187–192
65. Knapp RJ, Goldenberg R, Shock C, Cecil A, Watkins J, Miller C et al (2002) Antidepressant activity of memory-enhancing drugs in the reduction of submissive behavior model. Eur J Pharmacol 440:27–35
66. Szewczyk B, Poleszak E, Sowa-Kućma M, Wróbel A, Słotwiński S, Listos J, Wlaź P, Cichy A, Siwek A, Dybała M, Gołembiowska K, Pilc A, Nowak G (2009) The involvement of NMDA and AMPA receptors in the mechanism of antidepressant-like action of zinc in the forced swim test. Amino Acids [Epub ahead of print]
67. Duman RS, Heninger GR, Nestler EJ (1997) A molecular and cellular theory of depression. Arch Gen Psychiatry 54:597–606
68. Conti AC, Cryan JF, Dalvi A, Lucki I, Blendy JA (2002) cAMP response element-binding protein is essential for the upregulation of brain-derived neurotrophic factor transcription, but not the behavioral or endocrine responses to antidepressant drugs. J Neurosci 22:3262–3268
69. Miró X, Pérez-Torres S, Artigas F, Puigdomènech P, Palacios JM, Mengod G (2002) Regulation of cAMP phosphodiesterase mRNAs expression in rat brain by acute and chronic fluoxetine treatment. An in situ hybridization study. Neuropharmacology 43:1148–1157
70. Hayashi T, Umemori H, Mishina M, Yamamoto T (1999) The AMPA receptor interacts with and signals through the protein tyrosine kinase Lyn. Nature 397:72–76
71. Freed WJ, Dillon-Carter O, Kleinman JE (1993) Properties of [3H]AMPA binding in post-mortem human brain from psychotic subjects and controls: increases in caudate nucleus associated with suicide. Exp Neurol 121:48–56
72. Noga JT, Hyde TM, Herman MM, Spurney CF, Bigelow LB, Weinberger DR et al (1997) Glutamate receptors in the postmortem striatum of schizophrenic, suicide, and control brains. Synapse 27:168–176
73. de Kloet ER, Joels M, Holsboer F (2005) Stress and the brain: from adaptation to disease. Nat Rev Neurosci 6:463–475
74. Shors TJ, Thompson RF (1992) Acute stress impairs (or induces) synaptic long-term potentiation (LTP) but does not affect paired-pulse facilitation in the stratum radiatum of rat hippocampus. Synapse 11:262–265
75. Tocco G, Shors TJ, Baudry M, Thompson RF (1991) Selective increase of AMPA binding to the AMPA/quisqualate receptor in the hippocampus in response to acute stress. Brain Res 559:168–171
76. Karst H, Joels M (2003) Effect of chronic stress on synaptic currents in rat hippocampal dentate gyrus neurons. J Neurophysiol 89:625–633
77. Krugers HJ, Koolhaas JM, Bohus B, Korf J (1993) A single social stress-experience alters glutamate receptor-binding in rat hippocampal CA3 area. Neurosci Lett 154:73–77
78. Bartanusz V, Aubry JM, Pagliusi S, Jezova D, Baffi J, Kiss JZ (1995) Stress-induced changes in messenger RNA levels of N-methyl-D-aspartate and AMPA receptor subunits in selected regions of the rat hippocampus and hypothalamus. Neuroscience 66:247–252
79. Schwendt M, Jezova D (2000) Gene expression of two glutamate receptor subunits in response to repeated stress exposure in rat hippocampus. Cell Mol Neurobiol 20:319–329
80. Rosa ML, Guimaraes FS, Pearson RC, Del Bel EA (2002) Effects of single or repeated restraint stress on GluR1 and GluR2 flip and flop mRNA expression in the hippocampal formation. Brain Res Bull 59:117–124
81. Suenaga T, Morinobu S, Kawano K, Sawada T, Yamawaki S (2004) Influence of immobilization stress on the levels of CaMKII and phospho-CaMKII in the rat hippocampus. Int J Neuropsychopharmacol 7:299–309

82. Du J, Gray NA, Falke CA, Chen W, Yuan P, Szabo ST et al (2004) Modulation of synaptic plasticity by antimanic agents: the role of AMPA glutamate receptor subunit 1 synaptic expression. J Neurosci 24:6578–6589

83. Naylor P, Stewart CA, Wright SR, Pearson RC, Reid IC (1996) Repeated ECS induces GluR1 mRNA but not NMDAR1A-G mRNA in the rat hippocampus. Brain Res Mol Brain Res 35:349–353

84. Svenningsson P, Tzavara ET, Witkin JM, Fienberg AA, Nomikos GG, Greengard P (2002) Involvement of striatal and extrastriatal DARPP-32 in biochemical and behavioral effects of fluoxetine (Prozac). Proc Nat Acad Sci USA 99:3182–3187

85. Svenningsson P, Bateup H, Qi H, Takamiya K, Huganir RL, Spedding M, Roth BL, McEwen BS, Greengard P (2007) Involvement of AMPA receptor phosphorylation in antidepressant actions with special reference to tianeptine. Eur J Neurosci 26:3509–3517

86. Du J, Suzuki K, Wei Y, Wang Y, Blumenthal R, Chen Z, Falke C, Zarate CA Jr, Manji HK (2007) The anticonvulsants lamotrigine, riluzole, and valproate differentially regulate AMPA receptor membrane localization: relationship to clinical effects in mood disorders. Neuropsychopharmacology 32:793–802

87. Roche KW, O'Brien RJ, Mammen AL, Bernhardt J, Huganir RL (1996) Characterization of multiple phosphorylation sites on the AMPA receptor GluR1 subunit. Neuron 16:1179–1188

88. Barria A, Muller D, Derkach V, Griffith LC, Soderling TR (1997) Regulatory phosphorylation of AMPA-type glutamate receptors by CaM-KII during long-term potentiation. Science 276:2042–2045

89. Lee HK, Barbarosie M, Kameyama K, Bear MF, Huganir RL (2000) Regulation of distinct AMPA receptor phosphorylation sites during bidirectional synaptic plasticity. Nature 405:955–959

90. Du J, Gray NA, Falke C, Yuan P, Szabo S, Manji HK (2003) Structurally dissimilar antimanic agents modulate synaptic plasticity by regulating AMPA glutamate receptor subunit GluR1 synaptic expression. Ann N Y Acad Sci 1003:378–380

91. Brennan BP, Hudson JI, Jensen JE, McCarthy J, Roberts JL, Prescot AP, Cohen BM, Pope HG Jr, Renshaw PF, Ongür D (2010) Rapid enhancement of glutamatergic neurotransmission in bipolar depression following treatment with riluzole. Neuropsychopharmacology 35:834–846

92. Zarate CA Jr, Manji HK (2008) The role of AMPA receptor modulation in the treatment of neuropsychiatric diseases. Exp Neurol 211:7–10

93. Karasawa J, Shimazaki T, Kawashima N, Chaki S (2005) AMPA receptor stimulation mediates the antidepressant-like effect of a group II metabotropic glutamate receptor antagonist. Brain Res 1042:92–98

94. Maeng S, Zarate CA Jr, Du J, Schloesser RJ, McCammon J, Chen G, Manji HK (2008) Cellular mechanisms underlying the antidepressant effects of ketamine: role of alpha-amino-3-hydroxy-5-methylisoxazole-4-propionic acid receptors. Biol Psychiatry 63:349–352

95. Dybała M, Siwek A, Poleszak E, Pilc A, Nowak G (2008) Lack of NMDA-AMPA interaction in antidepressant-like effect of CGP 37849, an antagonist of NMDA receptor, in the forced swim test. J Neural Transm 115:1519–1520

96. Gould TD, O'Donnell KC, Dow ER, Du J, Chen G, Manji HK (2007) Involvement of AMPA receptors in the antidepressant-like effects of lithium in the mouse tail suspension test and forced swim test. Neuropharmacology 54:577–587

Glutamatergic Modulators for the Treatment of Major Depressive Disorder: Metabotropic Glutamate Receptors

Jeffrey M. Witkin

Abstract The majority of antidepressants facilitate monoamine neurotransmission within the central nervous system as their initial neurochemical processing event. These medicines do not fully serve patient needs in terms of response or remission, and suffer from some compliance issues. Among the various approaches that are being taken to discover improved therapeutics for major depressive disorders has been the targeting of the metabotropic glutamate receptors. Three major classes of mGlu receptors with a total of eight subtypes (mGlu1–8) have been considered. The bulk of data at present provides compelling evidence that mGlu5 and mGlu2/3 receptors are principal protein targets for drug discovery. These mGlu receptors have localization within primary mood circuits and are modulated by antidepressant treatment. Moreover, pharmacological blockade of mGlu5 (with MPEP, MTEP) or mGlu2/3 (with LY341495, MGS0039) receptors is associated with antidepressant-like biochemical and neurochemical changes in the mammalian brain. Small molecules that target these sites have been identified that display corresponding antidepressant-like behavioral effects. Agonist actions at mGlu2/3 receptors, demonstrated to possess anxiolytic activity in humans, might also be capable of impacting the progression of disease symptoms. Importantly, modulation of the mGlu5 and mGlu2/3 receptors impact ionotropic glutamate receptor signaling (NMDA, AMPA) that independently has been associated with mood disorders. The lack of biochemical and behavioral evidence linking other mGlu receptors to mood disorders is generally due to the absence of an appropriate arsenal of tools (e.g., selective, systemically-available compounds). However, there are data suggesting that mGlu1, mGlu7, and mGlu8 receptors might also play important roles in controlling mood. In contrast, mGlu6 receptors, with localization principally confined to retinal cells, are not likely candidate proteins. The patent and scientific literature suggests that the evaluation of these hypotheses in the clinic might be

J.M. Witkin
Neuroscience Discovery, Lilly Research Laboratories, Psychiatric Discovery, Lilly Corporate Center, Eli Lilly and Company, Indianapolis, IN, USA
e-mail: jwitkin@lilly.com

P. Skolnick (ed.), *Glutamate-based Therapies for Psychiatric Disorders*,
Milestones in Drug Therapy, DOI 10.1007/978-3-0346-0241-9_4,
© Springer Basel AG 2010

possible in the near future. With these awaited studies is the hope for new medicines
for the devastating and life-threatening diseases of mood that affect large segments
of the world community.

Abbreviations

ACPD	1-aminocyclopentane-trans-1,3R-dicarboxylic acid
ACPT-I	(1S,3R,4S)-1-aminocyclo-pentane-1,3,4-tricarboxylic acid
AMN082	N,N'-dibenzyhydryl-ethane-1,2-diamine dihydrochloride
AMPA	α-amino-3-hydroxy-5-methyl-4-isoxazolepropionic acid
BDNF	Brain-derived neurotrophic factor
cAMP	Adenosine $3',5'$-cyclic monophosphate
CNS	Central nervous system
CPPG	(RS)-alpha-cyclopropyl-4-phosphonophenyl glycine
DCG-IV	((2S,2'R,3'R)-2-(2',3'-dicarboxycyclopropyl)glycine)
ECT	Electroconvulsive treatment
EMQMCM	(3-ethyl-2-methyl-quinolin-6-yl)-(4-methoxy-cyclohexyl)-methanone methanesulfonate
FST	Forced swim test
L-CCG-I	(2S,1'S,2'S)-2-(carboxycyclopropyl)glycine
LY341495	9H-xanthene-9-propanoic acid, α-amino-α-[(1S,2S)-2-carboxycyclopropyl]-(αS)-(9CI)
mGlu	Metabotropic glutamate
MPEP	2-methyl-6-(phenylethynyl)pyridine
MSG0039	2-amino-3-[(3,4-dichlorophenyl)methoxy]-6-fluoro-(1R,2R,3R,5R,6R)-(9CI)
MTEP	3-[(2-methyl-1,3-thiazol-4-yl)ethynyl]pyridine
NBQX	2,3-dihydroxy-6-nitro-7-sulfamoylbenzo(f)quinoxaline
NMDA	N-methyl-D-aspartate
PHCCC	N-phenyl-7-(hydroxyimino)cyclopropa[b]chromen-1acarboxamide
(R,S)-PPG	(RS)-4-phosphonophenylglycine
SSRI	Selective serotonin reuptake inhibitors
TST	Tail suspension test

1 Introduction

Disorders of mood including major depressive disorder (MDD) and bipolar depression negatively and grossly impact the lives of millions of patients worldwide. Available medicines primarily consist of the monoamine uptake inhibitors with a general focus on the selective serotonin uptake inhibitors. Although these agents are generally effective and well-tolerated, there is a huge need for improved medical treatments. Improvements in both efficacy and side effects are required to control response and remission rates and some side effects that limit needed patient compliance. Some of the major new advances that point to the potential for improved future therapeutic options include dual and triple monoamine uptake inhibitors, nonconventional antidepressants such as tianeptine, and a number of augmentation strategies. In addition, exploration is underway with a number of mechanisms that might yield the next breakthrough in pharmacological therapy. These include endocannabiniod mechanisms, a host of natural products, neuropeptide systems such as galanin and melanin-concentrating hormone, growth and neurtrophic factors such as BDNF, VGF, and FGF and epigenetic mechanisms involving histone modification. A large area of current investigation is directed at protein targets involved in glutamate neurotransmission that include NMDA, AMPA, metabotropic glutamate receptors, and glia and associated excitatory amino acid transporters [1]. The present review will focus on the potential value mGlu receptors as novel protein targets for drug discovery.

2 Glutamate Modulation

Glutamate projections map to and from major brain areas that focus on mood, reaction to stress, emotional memory, and cognitive interpretation [2]. In addition to neuroanatomical rationale, convergent evidence from antidepressant actions, stress reactivity, brain imaging, and behavioral pharmacology has all implicated glutamatergic neurotransmissions as a major neurochemical system as a key regulator of mood [3, 4]. Research in this area has focused on NMDA receptors [5], AMPA receptors [6, 7], metabotropic glutamate receptors [8, 9], downstream signaling pathways [10], as well as on EAATs, glia, and glutamate/glutamine cycling [11]. The present review will concentrate on the potential role of the mGlu receptors in mood disorders and the manner in which these receptors could be pharmacologically enhanced or attenuated in order to bring about more favorable responses to those in need.

3 Metabotropic Glutamate Receptors

mGlu receptors are classified into three major classes with a total of eight subtypes (mGlu1–8) of which some splice variants exist. Metabotropic glutamate (mGlu) receptors function to regulate glutamate neuronal transmission by modulating the release of neurotransmitters and the postsynaptic responses to glutamate by way of downstream actions and presynaptic modulation impacts other neurochemical systems known to be involved in the regulation of mood [12]. A reasonable body of data exists to document that mGlu receptors have dynamic communication links with monoamine neurotransmission and with ionotropic glutamate receptors thought to control mood [8, 9, 13]. mGlu receptors are intimately linked to mood and antidepressant drug action and are actively being investigated as novel targets for antidepressant drug discovery efforts for at least the following reasons that will be summarized in the present review (1) pharmacological modulation of mGlu receptors can facilitate neuronal stem cell proliferation (neurogenesis) and the release of neurotransmitters that are associated with antidepressant efficacy; (2) localization of mGluRs in mood-disorder related neural circuits; (3) alterations in mGlu receptors with antidepressant treatments; and (4) antidepressant-like biochemical and behavioral control by mGlu receptors.

4 Group I mGlu Receptors

The group I mGlu receptors include the mGlu1 and mGlu5 receptors with splice variants (mGlu1α, β and mGlu5a,b) [14]. The distribution of the group I receptors overlaps with brain areas implicated in mood disorders [2, 14]. The biochemical coupling and localization of these receptors have been described (Table 1). Additional evidence supporting the involvement of Group I mGlu receptors in mood disorders is summarized below.

4.1 Antidepressants Modify mGlu Receptors

Subchronic administration of antidepressants decreased sensitivity of group I mGlu receptors. For example, subchronic antidepressant or ECT administration inhibited both ibotenate-induced cAMP accumulation and the interaction between ibotenate and noradrenaline [15]. Group I mGlu receptor activation by ACPD or DHPG, which caused an increase in the activity of neurons from the CA1 region of the hippocampus in rats, was inhibited by both multiple imipramine administration or ECT sessions [16]. Immunoreactivity of mGlu1α receptors was increased after multiple but not single ECT in the CA regions of the hippocampus [17, 18]. Olfactory bulbectomy is used as a model against which to assess antidepressant treatments. Lesioned rats demonstrated increases in hippocampal mGlu1a receptors; this increase was blocked by 14 days of amitriptyline dosing [19]. Likewise, expression of mGluR5a was increased with subchronic imipramine into CA1 and increased in CA3 after chronic ECT [18]. This receptor upregulation might be a compensatory mechanism developing after antidepressant treatment that results in receptor subsensitivity [17, 18]. In rats, chronic mild stress that can engender depressive-like symptoms [20] increased mGlu5 receptor protein in hippocampal CA1 [21].

Table 1 Anatomical, biochemical, and pharmacology relationship of mGlu receptors and mood[a]

Receptor	Coupling	Localizations and functions	Antidepressant-like biochemical effects[b]	Antidepressant-like behavioral effects[b]
Group I mGlu$_1$. Splice variants: mGlu1α, β	Gq coupled to increase in PLC; AC activation also noted [55]	Postsynaptic on neurons of cerebellum, olfactory bulb, CA3 region of hippocampus, thalamus, dentate gyrus, substantia nigra, and medial central gray [14] Group I mGlu receptors are also localized around iGlu receptors [69] Positive modulation of glutamate neurotransmission	Neurotrophin induction	EMQMCM and JNJ16567083 active in rat FST
mGlu$_5$ Splice variants: mGlu5a, b	Gq coupled to increase in PLC; AC activation also noted [55]	Postsynaptic on neurons and glia of telencephalic regions, CA1 and CA3 regions of the hippocampus, in the septum, basal ganglia, striataum, amygdala and nucleus accumbens [14] Group I mGlu receptors are also localized around iGlu receptors [69] Positive modulation of glutamate neurotransmission Expression of mGlu5 receptors on GABAergic interneurons in the cortex and hippocampus [70]; stimulation can increase in glutamate release [56] Activation of group I mGlu receptors leads to activation of NMDA receptors, mGlu5 receptors potentiate	CHPG did not alter GluR1 S845-phophorylation	Active in FST, TST (MPEP, MTEP) Active in OB (MTEP) Synergy with IMI in FST KO mouse – antidepressant-like phenotype; not responsive to MPEP in FST

(continued)

Table 1 (continued)

Receptor	Coupling	Localizations and functions	Antidepressant-like biochemical effects[b]	Antidepressant-like behavioral effects[b]
		NMDA-evoked current via G proteins, while mGlu1 receptors act via activation of Src tyrosine kinase [71]		
Group II			Neurogenesis with antagonists Neuroprotection with agonists Agonists – cFos activation more in anxiety than mood-regulation brain structures Agonists and antagonists increase cortical dopamine and serotonin	LY341495 active in rodent models MGS0039 active in rodent models
mGlu$_2$	Gi, Go coupled to decreases in AC	Neuronal localizations in accessory and external regions of the anterior olfactory bulb, pyramidal neurons in the enthorhinal and parasubicular cortical regions, and granule cells of the dentate gyrus [72] Principally localized presynaptically as an autoreceptor or heteroreceptor, and are seen in preterminal rather than terminal portions of axons [74] Activation negatively modulate glutamate and GABA neuronal transmission	See also Group II Antidepressant-like effect of mGlu2 −/− mice in FST Increased dopamine and glutamate release in mGlu2 −/− mice	See Group II
mGlu$_3$	Gi, Go coupled to decreases in AC	Primary localization postsynaptic to neurons in cerebral cortex and the caudate-putamen and in granule cells of the dentate gyrus and in glial cells [73]	See Group II	See Group II

		Activation negatively modulate glutamate and GABA neuronal transmission		
Group III				
mGlu₄	Gi, Go coupled to decreases in AC	Localizations presynaptic and postsynaptic in neurons. High receptor levels noted in rat cerebellum and granual cells of cerebellum		
		Also located in regions of cerebral cortex, in basal ganglia, and hippocampus [14]		
		Negative modulation of glutamate and GABA neuronal transmission suggested		
mGlu₆	Gi, Go coupled to decreases in AC	Highly localized in retinal tissues to dendrites of ON bipolar cells, with very low expression noted in brain tissue		
		Pharmacological, knockout, and human genetic data (night blindness) suggest primary role in the processing of visual sensory information		
MGlu₇ Splice variants: Mglu7a, B	Gi, Go coupled to decreases in AC	Localizations presynaptic and postsynaptic in neurons. Targeted presynaptically to the active zone of glutamate release	mGlu7 KO mice altered HPA axis sensitivity and increased BDNF levels	mGlu7 −/− mice demonstrated anxioltyic and antidepressant-like phenotypes
		Highly expressed in most brain areas, including neocortical regions, cingulate and piriform cortices, CA1, CA3 and DG regions of hippocampus, amygdala, locus		Antidepressant-like effects of AMN082 in WT but not mGlu7 −/− mice

(*continued*)

Table 1 (continued)

Receptor	Coupling	Localizations and functions	Antidepressant-like biochemical effects[b]	Antidepressant-like behavioral effects[b]
		coeruleus, and hypothalamic and thalamic nuclei [14]		
		Negative modulation of glutamate and GABA neuronal transmission suggested		
mGlu$_8$ Splice variants: mGlu8a, b	Gi, Go coupled to decreases in AC	Localizations pre and postsynaptic in neurons	cFos activation in stress areas with agonist DCPG	Anxiolytic-like effects of agonists sometimes reported
		Dominate in presynaptic terminals in the olfactory bulbs, piriform cortex, entorhinal cortex, hippocampus, and cerebellum [14]	Enhanced c-FOS activation to stress in mGlu8 KO	
		Negative modulation of glutamate (and possibly GABA neuronal transmission)		

Abbreviations: *AC* adenylyl cyclase, *CMS* chronic mild stress, *ECS* electroconvulsive shock, *FST* forced-swim test, *IMI* imipramine, *KO* receptor knockout, *OB* olfactory bulbetomized rat model, *PLC* phospholipase C, *PI* phosphoinositol, *ND* no data, *NR* no response, *TST* tail-suspension test

[a]See text for details and discussion of table entries

[b]See text for citations and discussion of table entries

4.2 Antidepressant-Like Behavioral Effects of Antagonists

Table 1 summarizes the major findings in this area with rodent models that detect antidepressant activity. The mGlu5 antagonists MPEP and MTEP decreased immobility in the tail-suspension test (TST, an antidepressant-like effect) in C57BL/6J mice [22–25]. In rats, MPEP was not active in the forced-swim test (FST) [22] but was active in mice [26] and MTEP was active in rats in a modified version of the FST [24]. Antidepressant- and anxioltyic-like effects in rodents were found with subchronic dosing with MPEP [27]. In the mouse FST, MPEP enhanced the effects of the tricyclic antidepressant imipramine without modifying brain drug levels [26]. MPEP (14 days) attenuated lesion-induced deficits in passive avoidance learning [28] and subchronic administration of MTEP attenuated the hyperactivity of olfactory bulbectomized rats[23]. mGlu5 receptor knockout mice displayed an antidepressant-like behavioral phenotype (decrease in the immobility compared to wildtype mice) [26]. In the mGlu5 knockout mice, imipramine further decreased the immobility time, while MPEP was not effective demonstrating that the effects of MPEP in the FST are due to mGlu5 receptor blockade [26].

Less is known about the antidepressant-like effects of mGlu1 receptor modulation due to the paucity of good pharmacological tools. EMQMCM was active in both the TST and in the modified FST in rats [24].

SSRI antidepressants display anxiolytic activity in patients with an onset of action generally comparable to their antidepressant effects. The Group I mGlu receptor antagonists have shown robust antianxiety activity in animal models [28].

5 Group II mGlu Receptors

The group II mGlu receptors (mGlu2 and mGlu3 receptors) are localized in moodrelevant brain structures [14] and transduce antidepressant-like neurochemical changes (Table 1). Given the lack of selective ligands for mGlu2 vs. mGlu3 receptors and the current lack of data, specific roles for these two subtypes of the group II mGlu receptors are difficult to assign at this time.

5.1 Antidepressants Modify mGlu Receptors

Although imipramine (21 days) did not alter mGlu2/3 receptor number, it reduced the sensitivity of these receptors to the inhibition of forskolin-stimulated cAMP accumulation in the rat brain cortical slices induced by the group II mGlu receptor agonist, 2R,4R-APDC [29]. A similar lack of receptor density change was observed in Rhesus monkey brain after 39 weeks of fluoxetine dosing [30]. Subchronic imipramine reduced mGlu2/3 receptor agonist-mediated inhibition of forskolinstimulated cAMP formation, while it enhanced mGlu2/3 receptor-mediated phosphoinositol responses in hippocampal slices [31]. It was hypothesized that imipramine reduces the function of presynaptic mGlu2/3 receptors, whereas the same manipulation enhances postsynaptic mGlu2/3 receptor function, which might contribute to increased glutamatergic synaptic transmission. When imipramine was combined with LY341495 or a low dose of the mGlu2/3 agonist LY379268, neuroadaptation to imipramine (change in $\beta 1$ adrenoceptor expression) occurred at shorter times than with imipramine alone [32], suggesting the possibility that mGlu2/3 receptor ligands might shorten the time required to exert full antidepressant effects, which is generally delayed by several weeks [33]. Matrisciano et al. [34] also reported the enhancement of neurogenesis induction in vitro with LY379268 and fluoxetine. In olfactory-bulbectomized rats, levels of mGlu2/3 receptors were decreased in hippocampus, an effect that was attenuated by amitriptyline administration for 14 days [19]. Decreases in mGlu2/3 receptors were also observed in the depressed Flinders line of rats [34]. However, it was recently reported that mGlu2/3 receptors are increased in the prefrontal cortex of depressed patients [30].

5.2 Antidepressant-Like Neurochemical Effects

Table 1 summarizes some of the key findings. Administration of the mGlu2/3 antagonists MGS0039 and LY341495 increased firing rates of dorsal raphe neurons and increased extracellular levels of serotonin in the medial prefrontal cortex (mPFC) [35]. The enhanced serotonin efflux in the mPFC by MGS0039 was attenuated by NBQX, an AMPA receptor antagonist [36]. Increases in serotonin and noradrenaline were also observed in the ventral hippocampus after systemic administration of LY341495 [8]. Dopamine efflux in the nucleus accumbens is related to mood/hedonic valuation. Direct application of MGS0039 into the nucleus accumbens shell increased dopamine release in this brain area, while local injection of the mGlu2/3 agonist LY354740 decreased release [37]. mGlu2 receptor $-/-$ mice displayed enhanced dopamine release in the nucleus accumbens following cocaine administration [38]. Emerging lines of evidence suggest that increased hippocampal neurogenesis might be a common mechanism of antidepressant treatments [6, 39, 40]. MGS0039 (for 14 days) increased progenitor cell proliferation in the dentate gyrus [41].

5.3 Antidepressant-Like Behavioral Effects of Antagonists

Table 1 summarizes some of the key findings in this area. MGS0039 and LY341495 engender antidepressant-like effects in behavioral despair models such as the FST and TST [8, 42, 43]. Antidepressant-like effects in the TST were also observed after MGS0039 for 5 days with no indication of tolerance. MGS0039 and LY341495 increased swimming behavior without changing climbing behavior in the rat FST, as observed with fluvoxamine and other selective serotonin reuptake inhibitors (SSRIs) [44]. LY341495 attenuated reward deficits after nicotine withdrawal in nicotine-dependent rats as evaluated by intracranial self-stimulation [45, 46]; thus, blockade of mGlu2/3 receptors opposes anhedonic effects. MGS0039 had antidepressant-like effects in the learned helplessness model; MGS0039 dosed for 7 days reduced escape failures without changing body weight gain, while imipramine showed the effects at the dose which markedly reduced body weight gain [47].

An antidepressant-like behavioral phenotype was reported in mice lacking mGlu2 receptors. In the FST, mGlu2 receptor $-/-$ mice were significantly more mobile compared to wild-type mice on the second day of testing, although there was no difference in immobility in the TST [38]. mGlu2 receptor $-/-$ knockout mice also showed an increase in locomotor sensitivity and conditioned place preference in association with repeated cocaine administration, indicating increased reinforcing effects. These later findings are consistent with the reported increase in dopamine efflux produced by MGS0039 in the nucleus accumbens shell as noted above [37].

mGlu2/3 receptor antagonists also have other in vivo effects that would support antidepressant therapy. MGS0039 displayed anxiolytic-like effects in some anxiety-relevant rodent models [13, 47]. Agonists of mGlu2/3 receptor agonists also exhibit anxiolytic effects in rodents and humans [48]. mGlu2/3 receptor antagonists also have procognitive effects [13] and drug-dependence therapeutic-like actions [46].

6   Group III mGlu Receptors

Group III mGlu receptors are classified into four receptor subtypes (mGlu4, mGlu6, mGlu7, and mGlu8 receptors) with splice variants. The localization and biochemical impact of these receptors is summarized in Table 1.

6.1   Antidepressants Modify mGlu Receptors

mGluR4a-immunoreactivity was not changed in rat brain after subchronic doses of imipramine [31, 49]. In contrast, mGluR7a-immunoreactivity was decreased after subchronic citalopram, but not imipramine treatment, in hippocampus and cortex [49]. A comparable dissociation of mGlu4 and 7 receptors was observed in olfactory-bulbectomized rats, where the decrease in mGlu7 receptors was prevented by 14-day dosing with amitriptyline whereas mGlu4 receptor decreases were not [19]. However, subchronic dosing with antidepressants failed to alter the effects of the group III mGlu receptor agonist, ACPT-1 on forskolin-stimulated cAMP accumulation [49].

6.2   Antidepressant-Like Behavioral Effects

Subtype specific and systemically-bioavailable pharmacological tools are lacking for the group III mGlu receptors. Antidepressant-like behavioral effects of these receptors has principally been based upon either central administration of compounds or on observations of the behavior of receptor knockout mice. The selective group III mGlu receptor agonist ACPT-I (icv) produced activity in the FST, an effect prevented by the group III mGlu receptor antagonist, CPPG [50, 51]. Klak et al. [52] confirmed activity of ACPT-I in the rat FST, an effect absent in the TST upon intraperitoneal dosing despite notable anxiolytic effects [53]. They also demonstrated that PHCCC (icv dosing), a positive allosteric modulator of mGluR4a receptors, given in combination with a noneffective dose of ACPT-I produced antidepressant-like effects in this assay. The specificity of the interaction was defined by the observation that the effect was blocked by CPPG [52].

RS-PPG (icv), an mGlu8 receptor agonist, produced a dose-dependent antidepressant-like effect in the rat FST [51]. mGlu7 receptor-deficient mice exhibited an antidepressant-like phenotype with shortened immobility times in the FST and TST [52]. AMN082, an agonist at mGlu7 receptors produced antidepressant-like effects in the mouse FST; this effect was absent in mGlu7 knockout mice [54]. A reduction in glutamate overflow, which has been observed after activation of group III mGlu receptors in several brain areas [12], might be responsible for the antidepressant-like effects of group III mGlu receptor agonists.

Group III mGlu receptor agonists also have shown anxiolytic activity in animal models [28].

7   mGlu Receptor Involvement in Mood Disorders

Based upon the limited data on the involvement of mGlu receptors in MDD and the absence of clinical data, a rudimentary model is proposed to account for the interactions of mGlu and other receptors and biological transduction processes (Fig. 1). A specific mapping of these mechanisms to the brain circuitry known at this time for mood disorders has not been previously proposed. However, as can be seen in Table 1, mGlu receptors are localized to brain structures that are known to be critical to depression [2].

7.1   Group I mGlu Receptors

mGlu receptors modulate the transmission of glutamate and of other neurotransmitters and involve multiple biochemical transduction pathways that have been previously shown to be involved in the regulation of mood. The summary of the literature above provides a foundation for appreciating the role of the metabotropic glutamate receptors in mood disorders and as potential targets for small molecule discovery efforts. Group I mGlu receptors could produce their antidepressant-like effects in various ways. Although the majority of group I mGlu receptors are located postsynaptically [55], a presynaptic localization of group I mGlu receptors has also been reported [56] and the reduction of glutamate release might be relevant

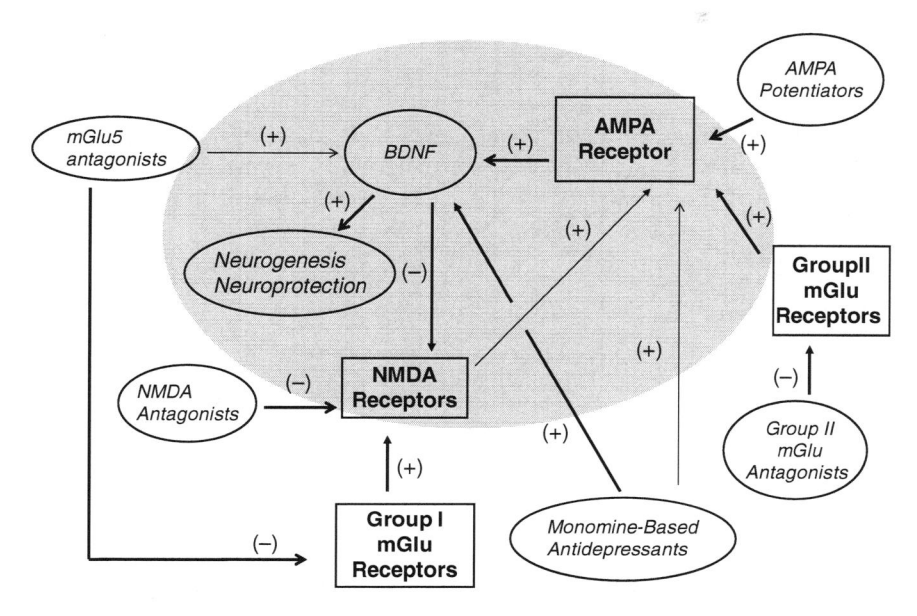

Fig. 1 A model for the induction of antidepressant-like effects by mGlu receptors. *Ovals* represent compounds, drugs, or chemical entities that can interact with receptors (*squares*). *Plus* signs represent biological interactions that are excitatory whereas *minus* signs represent biological interactions that are inhibitory. The *shaded oval* represents the key elements thought to be involved in initial transduction of antidepressant-related biochemistry. As with monoamine-based antidepressants, the mGlu receptors likely impact antidepressant-like biological responses through the dampening of NMDA receptor function and the amplification of AMPA receptor function. Data supporting this model are in the text and predominantly outlined in Sect. 5. The model is uncertain in a number of places due to limited data (*nonbold arrows*). There is very limited information on the specific mechanisms responsible for antidepressant-like effects of the group III mGlu receptors and therefore this class of mGlu receptors is not represented in the model. Additional details of the mechanisms (e.g., secondary biochemical pathways) contributing to these biochemical changes are not shown for clarity at this level. Modified from from [9] with permission from Elsevier

to the anxiolytic/antidepressant actions of mGlu5 antagonists [57]. In addition, mGlu5 receptor antagonists dampen NMDA receptor message [58] and function in key brain areas [59]. Therefore, mGlu5 receptor blockade might utilize NMDA receptor antagonism as a final common pathway for the anxiolytic- and antidepressant-like actions transduced (see [5]).

The group I mGlu receptor antagonists could also impact mood through their effects on neurotrophic factors. The generation of BDNF by antidepressants and its downstream enhancement of neural cell integrity and neurogenesis has become a heuristic model of antidepressant drug action [6, 7]. Subchronic administration of MPEP or the tricyclic antidepressant desipramine increased BDNF mRNA level in the rat hippocampus [60]. These data along with the findings that BDNF can down regulate NMDA receptor function [61] provide another potential route to the antidepressant-like efficacy of group I mGlu receptor antagonists.

A cooperative interaction of mGlu5 receptors with monoaminergic systems is another potential antidepressant-relevant pathway to consider. MPEP produces neuroendocrine responses in common with monoamine-based antidepressants: an increase in plasma corticosterone after acute dosing (this effect was attenuated by a 5-HT_{1A} receptor antagonist) and desensitization of neuroendocrine responses stimulated by a 5-HT_{1A} agonist after 5 days of dosing [62]. Interactions of MPEP and MTEP with noradrenaline secretion have also been described [63].

7.2 Group II mGlu Receptors

The current hypothesis underlying the mechanisms by which mGlu2/3 receptor antagonism engenders antidepressant-like behavioral and neurochemical effects involves the activation of AMPA receptors resulting from increased glutamate efflux. This hypothesis is supported by the data showing that both neurochemical and behavioral effects of mGlu2/3 receptor antagonists can be prevented by AMPA receptor antagonists as described above (see also [7]) and with the biochemical and behavioral data supporting an important role for AMPA receptor potentiation in mood disorders [6, 7].

7.3 Group III mGlu Receptors

As with group I mGlu receptor antagonists, reduced glutamate release might account for the biological effects of the group III mGlu receptor agonists [12]. Clinically used antidepressants reduce glutamate release both in vivo and in vitro and after acute and/or subchronic treatment with several antidepressants [64–66]. Nonetheless, other actions transduced by group III mGlu receptor activation such as inhibition of GABAergic neurotransmission or control of serotonin release [67, 68] cannot be ruled out.

Acknowledgments I thank the following colleagues for previous discussions that have helped shape some of the thoughts expressed here: Andrew J. Alt, David Bleakman, Shigeyuki Chaki, Gerard J. Marek, Andrzej Pilc, Gabriel Nowak, Darryle D. Schoepp, and Phil Skolnick.

References

1. Witkin JM, Li X (2009) New approaches to the management of major depressive disorder. In: Enna SJ, Williams M (eds) Contemporary aspects of biomedical research: biomedical issues. Advances in pharmacology, vol 57. Academic, New York, pp 347–379
2. Price JL, Drevets WC (2010) Neurocircuitry of mood disorders. Neuropsychopharmacology 35:192–216

3. Hashimoto K (2009) Emerging role of glutamate in the pathophysiology of major depressive disorder. Brain Res Rev 61:105–123
4. Schoepp DD, Skolnick P (2010) Introduction. In: Skolnick P (ed) Glutamate-based therapies for psychiatric disorders. Springer, Basel
5. Skolnick P, Popik P, Trullas R (2010) N-Methyl-D-Aspartate (NMDA) antagonists for the treatment of depression. In: Skolnick P (ed) Glutamate-based therapies for psychiatric disorders. Springer, Basel
6. Alt A, Nisenbaum ES, Bleakman D, Witkin JM (2006) A role for AMPA receptors in mood disorders. Biochem Pharmacol 71:1273–1288
7. Nisenbaum E, Witkin JM (2010) Positive allosteric modulation of AMPA receptors: a novel potential antidepressant therapy. In: Skolnick P (ed) Glutamate-based therapies for psychiatric disorders. Springer, Basel
8. Witkin JM, Marek GJ, Johnson BG, Schoepp DD (2007) Metabotropic glutamate receptors in the control of mood disorders. CNS Neurol Disord Drug Targets 6:87–100
9. Pilc A, Chaki S, Nowak G, Witkin JM (2008) Mood disorders: regulation by metabotropic glutamate receptors. Biochem Pharmacol 75:997–1006
10. Szabo ST, Machado-Vieira R, Yuan P, Wang Y, Wei Y, Falke C, Cirelli C, Tononi G, Manji HK, Du J (2009) Glutamate receptors as targets of protein kinase C in the pathophysiology and treatment of animal models of mania. Neuropharmacology 56:47–55
11. Valentine GW, Sanacora G (2009) Targeting glial physiology and glutamate cycling in the treatment of depression. Biochem Pharmacol 78:431–439
12. Cartmell J, Schoepp DD (2000) Regulation of neurotransmitter release by metabotropic glutamate receptors. J Neurochem 75:889–907
13. Witkin JM, Eiler WJA (2006) Antagonism of metabotropic glutamate group II receptors in the potential treatment of neurological and neuropsychiatric disorders. Drug Dev Res 67:757–769
14. Ferraguti F, Shigemoto R (2006) Metabotropic glutamate receptors. Cell Tissue Res 326:483–504
15. Pilc A, Legutko B (1995) The influence of prolonged antidepressant treatment on the changes in cyclic AMP accumulation induced by excitatory amino acids in rat cerebral cortical slices. NeuroReport 7:85–88
16. Pilc A, Branski P, Palucha A, Tokarski K, Bijak M (1998) Antidepressant treatment influences group I of glutamate metabotropic receptors in slices from hippocampal CA1 region. Eur J Pharmacol 349:83–87
17. Bajkowska M, Branski P, Smialowska M, Pilc A (1995) Effect of chronic antidepressant or electroconvulsive shock treatment on mGLuR1a immunoreactivity expression in the rat hippocampus. Pol J Pharmacol 51:539–541
18. Smialowska M, Szewczyk B, Branski P, Wieronska JM, Palucha A, Bajkowska M, Pilc A (2002) Effect of chronic imipramine or electroconvulsive shock on the expression of mGluR1a and mGluR5a immunoreactivity in rat brain hippocampus. Neuropharmacology 42:1016–1023
19. Wierońska JM, Legutko B, Dudys D, Pilc A (2008) Olfactory bulbectomy and amitriptyline treatment influences mGlu receptors expression in the mouse brain hippocampus. Pharmacol Rep 60:844–855
20. Papp M, Moryl E, Willner P (1999) Pharmacological validation of the chronic mild stress model of depression. Eur J Pharmacol 296:129–136
21. Wieronska JM, Branski P, Szewczyk B, Palucha A, Papp M, Gruca P, Moryl E, Pilc A (2001) Changes in the expression of metabotropic glutamate receptor 5 (mGluR5) in the rat hippocampus in an animal model of depression. Pol J Pharmacol 53:659–662
22. Tatarczynska E, Klodzinska A, Chojnacka-Wojcik E, Palucha A, Gasparini F, Kuhn R, Pilc A (2001) Potential anxiolytic- and antidepressant-like effects of MPEP, a potent, selective and systemically active mGlu5 receptor antagonist. Br J Pharmacol 132:1423–1430

23. Palucha A, Branski P, Szewczyk B, Wieronska JM, Klak K, Pilc A (2005) Potential antidepressant-like effect of MTEP, a potent and highly selective mGluR5 antagonist. Pharmacol Biochem Behav 81:901–906

24. Belozertseva IV, Kos T, Popik P, Danysz W, Bespalov AY (2007) Antidepressant-like effects of mGluR1 and mGluR5 antagonists in the rat forced swim and the mouse tail suspension tests. Eur Neuropsychopharmacol 17:172–179

25. Palucha A, Pilc A (2007) Metabotropic receptor ligands as possible anxiolytic and antidepressant agents. Pharmacol Ther 115:116–147

26. Li X, Need AB, Baez M, Witkin JM (2006) Metabotropic glutamate 5 receptor antagonism is associated with antidepressant-like effects in mice. J Pharmacol Exp Ther 319:254–259

27. Pilc A, Klodzinska A, Branski P, Nowak G, Palucha A, Szewczyk B, Tatarczynska E, Chojnacka-Wojcik E, Wieronska JM (2002) Multiple MPEP administrations evoke anxiolytic- and antidepressant-like effects in rats. Neuropharmacology 43:181–187

28. Pilc A et al (2010) Metabotropic approaches to anxiety. In: Skolnick P (ed) Glutamate-based therapies for psychiatric disorders. Springer, Basel

29. Pałucha A, Brański P, Kłak K, Sowa M (2007) Chronic imipramine treatment reduces inhibitory properties of group II mGlu receptors without affecting their density or affinity. Pharmacol Rep 59:525–530

30. Feyissa AM, Woolverton WL, Miguel-Hidalgo JJ, Wang Z, Kyle PB, Hasler G, Stockmeier CA, Iyo AH, Karolewicz B (2010) Elevated level of metabotropic glutamate receptor 2/3 in the prefrontal cortex in major depression. Prog Neuropsychopharmacol Biol Psychiatry 34 (2):279–283

31. Matrisciano F, Storto M, Ngomba RT, Cappuccio I, Caricasole A, Scaccianoce S, Riozzi B, Melchiorri D, Nicoletti F (2002) Imipramine treatment up-regulates the expression and function of mGlu2/3 metabotropic glutamate receptors in the rat hippocampus. Neuropharmacology 42:1008–1015

32. Matrisciano F, Scaccianoce S, Del Bianco P, Panaccione I, Canudas AM, Battaglia G, Riozzi B, Ngomba RT, Molinaro G, Tatarelli R et al (2005) Metabotropic glutamate receptors and neuroadaptation to antidepressants: imipramine-induced down-regulation of beta-adrenergic receptors in mice treated with metabotropic glutamate 2/3 receptor ligands. J Neurochem 93:1345–1352

33. Katz MM, Tekell JL, Bowden CL, Brannan S, Houston JP, Berman N et al (2004) Onset and early behavioral effects of pharmacologically different antidepressants and placebo in depression. Neuropsychopharmacology 29:566–579

34. Matrisciano F, Zusso M, Panaccione I, Turriziani B, Caruso A, Iacovelli L, Noviello L, Togna G, Melchiorri D, Debetto P et al (2008) Synergism between fluoxetine and the mGlu2/3 receptor agonist, LY379268, in an in vitro model for antidepressant drug-induced neurogenesis. Neuropharmacology 54:428–437

35. Kawashima N, Karasawa J, Shimazaki T, Chaki S, Okuyama S, Yasuhara A, Nakazato A (2005) Neuropharmacological profiles of antagonists of group II metabotropic glutamate receptors. Neurosci Lett 378:131–134

36. Karasawa J, Shimazaki T, Kawashima N, Chaki S (2005) AMPA receptor stimulation mediates the antidepressant-like effect of a group II metabotropic glutamate receptor antagonist. Brain Res 1042:92–98

37. Karasawa J, Yoshimizu T, Chaki S (2006) A metabotropic glutamate 2/3 receptor antagonist, MGS0039, increases extracellular dopamine levels in the nucleus accumbens shell. Neurosci Lett 393:127–130

38. Morishima Y, Miyakawa T, Furuyashiki T, Tanaka Y, Mizuma H, Nakanishi S (2005) Enhanced cocaine responsiveness and impaired motor coordination in metabotropic glutamate receptor subtype 2 knockout mice. Proc Natl Acad Sci 102:4170–4175

39. Santarelli L, Saxe M, Gross C, Surget A, Battaglia F, Dulawa S, Weisstaub N, Lee J, Duman R, Arancio O et al (2003) Requirement of hippocampal neurogenesis for the behavioral effects of antidepressants. Science 301:805–809

40. Duman RS, Nakagawa S, Malberg J (02001) Regulation of adult neurogenesis by antidepressant treatment. Neuropsychopharmacology 25:836–844

41. Yoshimizu T, Chaki S (2004) Increased cell proliferation in the adult mouse hippocampus following chronic administration of group II metabotropic glutamate receptor antagonist, MGS0039. Biochem Biophys Res Commun 315:493–496

42. Chaki S, Yoshikawa R, Hirota S, Shimazaki T, Maeda M, Kawashima N, Yoshimizu T, Yasuhara A, Sakagami K, Okuyama S et al (2004) MGS0039: a potent and selective group II metabotropic glutamate receptor antagonist with antidepressant-like activity. Neuropharmacology 46:457–467

43. Bespalov AY, van Gaalen MM, Sukhotina IA, Wicke K, Mezler M, Schoemaker H, Gross G (2008) Behavioral characterization of the mGlu group II/III receptor antagonist, LY-341495, in animal models of anxiety and depression. Eur J Pharmacol 592:96–102

44. Detke MJ, Rickels M, Lucki I (1995) Active behaviors in the rat forced swimming test differentially produced by serotonergic and noradrenergic antidepressants. Psychopharmacology 121:66–72

45. Kenny PJ, Gasparini F, Markou A (2003) Group II metabotropic and alpha-amino-3-hydroxy-5-methyl-4-isoxazole propionate (AMPA)/kainate glutamate receptors regulate the deficit in brain reward function associated with nicotine withdrawal in rats. J Pharmacol Exp Ther 306:1068–1076

46. Markou A, Semenova S (2010) Metabotropic glutamate receptors as targets for the treatment of drug and alcohol dependence. In: Skolnick P (ed) Glutamate-based therapies for psychiatric disorders. Springer, Basel

47. Yoshimizu T, Shimazaki T, Ito A, Chaki S (2006) An mGluR2/3 antagonist, MGS0039, exerts antidepressant and anxiolytic effects in behavioral models in rats. Psychopharmacology 186:587–593

48. Swanson CJ, Bures M, Johnson MP, Linden AM, Monn JA, Schoepp DD (2005) Metabotropic glutamate receptors as novel targets for anxiety and stress disorders. Nat Rev Drug Discov 4:131–144

49. Wierońska JM, Kłak K, Pałucha A, Brański P, Pilc A (2007) Citalopram influences mGlu7, but not mGlu4 receptors' expression in the rat brain hippocampus and cortex. Brain Res 1184:88–95

50. Tatarczynska E, Palucha A, Szewczyk B, Chojnacka-Wojcik E, Wieronska J, Pilc A (2002) Anxiolytic- and antidepressant-like effects of group III metabotropic glutamate agonist (1S, 3R, 4S)-1-aminocyclopentane-1, 3, 4-tricarboxylic acid (ACPT-I) in rats. Pol J Pharmacol 54:707–710

51. Palucha A, Tatarczynska E, Branski P, Szewczyk B, Wieronska JM, Klodzinska A, Chojnacka-Wojcik E, Nowak G, Pilc A (2004) Group III mGlu receptor agonists produce anxiolytic- and antidepressant-like effects after central administration in rats. Neuropharmacology 46:151–159

52. Klak K, Palucha A, Branski P, Sowa M, Pilc A (2007) Combined administration of PHCCC, a positive allosteric modulator of mGlu4 receptors and ACPT-I, mGlu III receptor agonist evokes antidepressant-like effects in rats. Amino Acids 32:169–172

53. Cryan JF, Kelly PH, Neijt HC, Sansig G, Flor PJ, van der Putten H (2003) Antidepressant and anxiolytic-like effects in mice lacking the group III metabotropic glutamate receptor mGluR7. Eur J Neurosci 17:2409–2417

54. Palucha A, Klak K, Branski P, Putten H, Flor P, Pilc A (2007) Activation of the mGlu7 receptor elicits antidepressant-like effects in mice. Psychopharmacology 194:555–562

55. Conn PJ, Pin JP (1997) Pharmacology and functions of metabotropic glutamate receptors. Annu Rev Pharmacol Toxicol 37:205–237

56. Moroni F, Cozzi A, Lombardi G, Sourtcheva S, Leonardi P, Carfi M, Pellicciari R (1998) Presynaptic mGlu(1) type receptors potentiate transmitter output in the rat cortex. Eur J Pharmacol 347:189–195

57. Thomas LS, Jane DE, Gasparini F, Croucher MJ (2001) Glutamate release inhibiting properties of the novel mGlu(5) receptor antagonist 2-methyl-6-(phenylethynyl)-pyridine (MPEP): complementary in vitro and in vivo evidence. Neuropharmacology 41:523–527

58. Cowen MS, Djouma E, Lawrence AJ (2005) The metabotropic glutamate 5 receptor antagonist 3-[(2-methyl-1, 3-thiazol-4-yl)ethynyl]-pyridine reduces ethanol self-administration in multiple strains of alcohol-preferring rats and regulates olfactory glutamatergic systems. J Pharmacol Exp Ther 315:590–600

59. Attucci S, Carla V, Mannaioni G, Moroni F (2001) Activation of type 5 metabotropic glutamate receptors enhances NMDA responses in mice cortical wedges. Br J Pharmacol 132:799–806

60. Legutko B, Szewczyk B, Pomierny-Chamiolo L, Nowak G, Pilc A (2006) Effect of MPEP treatment on brain-derived neurotrophic factor gene expression. Pharmacol Rep 58:427–430

61. Skolnick P (1999) Antidepressants for the new millennium. Eur J Pharmacol 375:31–40

62. Bradbury MJ, Giracello DR, Chapman DF, Holtz G, Schaffhauser H, Rao SP et al (2003) Metabotropic glutamate receptor 5 antagonist-induced stimulation of hypothalamic–pituitary–adrenal axis activity: interaction with serotonergic systems. Neuropharmacology 44:562–572

63. Page ME, Szeliga P, Gasparini F, Cryan JF (2005) Blockade of the mGlu5 receptor decreases basal and stress-induced cortical norepinephrine in rodents. Psychopharmacology 179:240–246

64. Prikhozhan AV, Kovalev GI, Raevskii KS (1990) Effects of antidepressive agents on glutamatergic autoregulatory presynaptic mechanism in the rat cerebral cortex. Biull Eksp Biol Med 110:624–626

65. Golembiowska K, Dziubina A (2000) Effect of acute and chronic administration of citalopram on glutamate and aspartate release in the rat prefrontal cortex. Pol J Pharmacol 52:441–448

66. Bonanno G, Giambelli R, Raiteri L, Tiraboschi E, Zappettini S, Musazzi L et al (2005) Chronic antidepressants reduce depolarization-evoked glutamate release and protein interactions favoring formation of SNARE complex in hippocampus. J Neurosci 25:3270–3279

67. Gereau RW, Conn PJ (1995) Multiple presynaptic metabotropic glutamate receptors modulate excitatory and inhibitory synaptic transmission in hippocampal area CA1. J Neurosci 15:6879–6889

68. Marek GJ, Wright RA, Schoepp DD, Monn JA, Aghajanian GK (2002) Physiological antagonism between 5-hydroxytryptamine(2A) and group II metabotropic glutamate receptors in prefrontal cortex. J Pharmacol Exp Ther 292:76–87

69. Lujan R, Nusser Z, Roberts JDB, Shigemoto R, Somogyi P (1996) Perisynaptic location of metabotropic glutamate receptors mGluR1 and mGluR5 on dendrites and dendritic spines in the rat hippocampus. Eur J Neurosci 8:1488–1500

70. Kerner JA, Standaert DG, Penney JB, Young AB, Landwehrmeyer GB (1997) Expression of group one metabotropic glutamate receptor subunit mRNAs in neurochemically identified neurons in the rat neostriatum, neocortex, and hippocampus. Mol Brain Res 48:259–269

71. Benquet P, Gee CE, Gerber U (2002) Two distinct signaling pathways upregulate NMDA receptor responses via two distinct metabotropic glutamate receptor subtypes. J Neurosci 22(22):9679–9686

72. Ohishi H, Shigemoto R, Nakanishi S, Mizuno N (1993a) Distribution of the messenger RNA for a metabotropic glutamate receptor, mGluR2, in the central nervous system of the rat. Neuroscience 53:1009–1018

73. Ohishi H, Shigemoto R, Nakanishi S, Mizuno N (1993b) Distribution of the mRNA for a metabotropic glutamate receptor (mGluR3) in the rat brain: an in situ hybridization study. J Comp Neurol 335:252–266

74. Shigemoto R, Kinoshita A, Wada E, Nomura S, Ohishi H, Takada M, Flor PJ, Neki A, Abe T, Nakanishi S, Mizuno N (1997) Differential presynaptic localization of metabotropic glutamate receptor subtypes in the rat hippocampus. J Neurosci 17:7503–7522

Positive Modulation of AMPA Receptors as a Broad-Spectrum Strategy for Treating Neuropsychiatric Disorders

Gary Lynch, Julie C. Lauterborn, and Christine M. Gall

Abstract The invention of centrally active, positive modulators of AMPA-type glutamate receptors ("ampakines") was prompted by the expectation that enhancing monosynaptic, fast excitatory post synaptic currents (EPSCs) would increase throughput in cortical networks and lower the threshold for induction of long-term potentiation (LTP), two events that could potentially enhance memory and cognition. Preclinical work has largely confirmed these various predictions and a small set of reports describe positive effects in humans. Later work raised the possibility that ampakines might have somewhat broader applications, a list that now extends from respiratory distress through autism, ADHD, depression, and schizophrenia. We here describe two general hypotheses to explain why ampakines might have therapeutic value with regard to disparate psychiatric illnesses. The first of these begins with the late nineteenth century idea that the cortex, in addition to its more traditionally understood functions, serves to regulate disturbances in lower brain systems. A version of this argument emerged in the 1950s as part of the intense research that followed on the discovery of links between the reticular formation in generating EEG and behavioral arousal. Various groups obtained evidence of a reticular–frontal cortical "loop" that served to modulate behavioral/physiological excitability. More recently, Carlsson and Carlsson expanded the loop concept to include brainstem biogenic amines; their model makes the explicit prediction that enhancing descending glutamatergic projections can be used to offset abnormal activity in the ascending dopaminergic system. Collectively, these theoretical positions point to the conclusion that positive modulation of cortical AMPA receptors should acutely reduce symptomology in disorders involving norepinephrine, dopamine, and serotonin. In accord with this assumption, ampakines depress high levels of arousal, counteract the effects of stimulants, and correct behavioral abnormalities in animal models of schizophrenia, ADHD, and

G. Lynch (✉)
Department of Psychiatry and Human Behavior, Gillespie Neuroscience Research Facility, University of California at Irvine, 837 Health Science Road, Irvine, CA 92697-4291, USA
e-mail: glynch@uci.edu

P. Skolnick (ed.), *Glutamate-based Therapies for Psychiatric Disorders*,
Milestones in Drug Therapy, DOI 10.1007/978-3-0346-0241-9_5,
© Springer Basel AG 2010

depression. Work on humans is limited although recent evidence points to clinically meaningful improvements in ADHD. The second hypothesis grows out of two literatures, one defining the machinery that reorganizes the spine cytoskeleton so as to encode memory and the other suggesting that defects in these same cellular processes contribute to a surprisingly large number of disorders involving memory and cognition. With regard to the present chapter, these observations are united by the discovery that a potent regulator of cytoskeleton reorganization (Brain-Derived Neurotrophic Factor: BDNF) is up-regulated by ampakines. The possibility thus arises that daily treatments with the drugs could be used to drive structural changes otherwise blocked by any of several genetic or behavioral conditions. Tests of this argument have so far been limited to animal models but the results are encouraging. Daily treatments with a short half-life ampakine that increases cortical BDNF concentrations restored synaptic plasticity in middle-aged rats and in mice carrying the Huntington's disease mutation. In the latter case, normalization of structural and physiological plasticity was accompanied by marked improvements in learning. Collectively, the results subsumed under the two hypotheses have to be viewed as remarkable: a class of drugs that has a single biophysical target produces positive changes of unprecedented breadth. But ampakines are still a recent invention arising from ideas in basic neuroscience. Whether the observed effects in animals actually relate to the two hypotheses remains to be formally tested and, above all else, there stands the absence of data from extensive clinical trials.

1   Introduction

AMPA-type glutamate receptors (AMPARs), so far as the point has been tested, mediate fast excitatory transmission throughout the central nervous system. Results from studies of hippocampus indicate that the receptors are central to two additional functions, the first of which involves plasticity. Stimulation of AMPARs provides the depolarization needed to unblock colocalized, voltage sensitive NMDA-type glutamate receptors [1] – this event results in the postsynaptic calcium pulse that sets in motion the multiple sequences leading to long-term potentiation (LTP) [2–4]. The second AMPAR-dependent effect is complex but possibly involved in the maintenance of neuronal viability. Specifically, there is now a sizeable collection of results demonstrating that excitatory input to cortical neurons drives the transcription of neuronal growth factors, including the critically important Brain-Derived Neurotrophic Factor (BDNF) [5, 6].

These points prompted the invention of the first compounds ("ampakines") that rapidly cross the blood–brain barrier and increase the size of monosynaptic, AMPAR-mediated EPSPs. As predicted, these early drugs facilitated LTP and accelerated a form of learning known to be dependent on hippocampus [7, 8]. Despite their effects on monosynaptic transmission and LTP, the agents caused no evident disturbances to baseline physiology or behavior. Subsequent work identified an ampakine binding pocket in the external domains of AMPARs and provided detailed descriptions of how the compounds affect receptor kinetics [9]. A plausible picture of drug action emerged when these discoveries were combined with information on which kinetic parameters terminate EPSCs at adult synapses. This material has been covered in several recent reviews [10–15].

The first behavioral work on ampakines focused on memory encoding. Two lines of reasoning led to the prediction that the compounds would produce an enhancing effect. First, facilitating fast EPSCs should increase the flow of activity through polysynaptic networks, an effect demonstrated in the hippocampus' tri-synaptic circuit [16], and thereby increase the cortex's ability to process complex information. Second, as noted, the enlarged synaptic depolarization caused by ampakines was expected to lower the threshold for opening NMDAR channels and thus promote the formation of LTP; the latter point has also been confirmed experimentally [17]. In accord with these arguments, ampakines markedly improved retention scores of rats in a radial maze [18]. It was then shown that the drugs reduce the amount of training needed to encode long-term memory in a two-odor discrimination task [7, 19] and in a conditioned fear paradigm [20]. Positive results for monkeys and humans were reported somewhat later [10, 21]. In all, a sizeable literature points to the conclusion that ampakines act as memory enhancers across species and tasks.

Later research along very different lines greatly expanded the potential uses for ampakines. An exciting and recent set of discoveries indicate that the drugs, possibly through a normalizing action on brainstem circuits that pace inhalation/exhalation, alleviate respiratory distress arising from opioid anesthesia in rats and humans [22–24] or as a component of Rett Syndrome as determined in a mouse model [25]. There is also evidence that repeated treatments reduce pathology in animal models of ischemia, Huntington's disease [26, 27], and Parkinson's disease [28–30]. These neuroprotective actions are thought to at least in part reflect ampakine-induced increases in neurotrophic factor expression. But the largest body of work outside of memory involves neuropsychiatric disorders, a topic that forms the subject matter of the present review.

2 Working Hypotheses

2.1 Cortical Control of Biogenic Amine Systems

We hypothesized that enhancing fast EPSCs in the cortical telencephalon would have positive effects in a diverse group of psychiatric disorders involving (a) biogenic amine neurotransmitters or (b) the spine cytoskeleton. The arguments for the first of these ideas can be traced back to the nineteenth century theories of Hughlings-Jackson and specifically the argument that "higher" levels of the brain inhibit activity in "lower" regions ([31] for review). Mental disorders can thus arise from a reduction in cortical control rather than from a primary defect at lower centers. Half a century later, an explosion of interest in the reticular formation sparked a renewed interest in cortical regulation of the lower brain and basic psychological variables [32]. Several groups were able to show that frontal fields regulate reticular activity and various signs of reflex excitability associated with it [33–35]. Interest in the area was, however, soon diverted into other avenues when Swedish researchers succeeded in mapping the distribution of the biogenic amine projections from the brainstem [36, 37]. Arousal, and much more, was then seen to involve multiple systems with diffuse reticular-like projections [38, 39]. Conceptually, there was no reason not to incorporate these discoveries into the reticulo–cortical feedback models, and indeed an early experiment showed that local lesions to the frontal pole of rat cortex greatly amplify the stimulant effects of amphetamine, a drug intimately related to biogenic amines [40].

Recently, other workers have formally proposed expanding the cortical feedback model to include the dopaminergic system and have used the results to develop a novel schizophrenia hypothesis. They argue that a frontal–striatal–nigral *descending* circuit serves to regulate output from the *ascending* dopamine system, and that hypofunctionality in the glutamatergic projections in the initial (cortex to ventral striatum) link leads to the hyperdopaminergic component of schizophrenia [41, 42]. The anatomical arrangements that provide the backbone of the argument [43] also apply to the serotonin and norepinephrine projections. For example, the ventral striatum massively innervates the habenula, which connects with both the raphe complex (serotonin) and the locus coeruleus (norepinephrine); direct connections from frontal cortex to the brainstem nuclei have also been described (Fig. 1) [44]. There is thus a possibility that the argument for a hypoglutamatergic cortical link in schizophrenia constitutes a special case of a general hypothesis regarding several psychiatric disorders involving biogenic amines.

It follows from the above that enhancing glutamatergic transmission within cortex and/or in descending cortical projections might serve to correct abnormal activity in ascending biogenic amine systems and thereby reduce symptomology in various disorders. This argument provided a rationale for exploring the use of ampakines in such conditions (see below).

2.2 Defects in the Spine Cytoskeleton May Be a Common Element in Psychiatric Disorders

Dendritic spine abnormalities are a characteristic feature of numerous disorders associated with cognitive impairments. Purpura was the first to report that cortical neurons in children with nonsyndromic retardation have excessive numbers of long, thin spines accompanied by a reduction in the frequency of stubby and mushroom-like spines [45]. Subsequent studies found similar spine abnormalities in Fragile-X [46–48], Down's [49], and Rett [50] syndromes; for a review see [51]. These observations led to the hypothesis that mental retardation emerges whenever

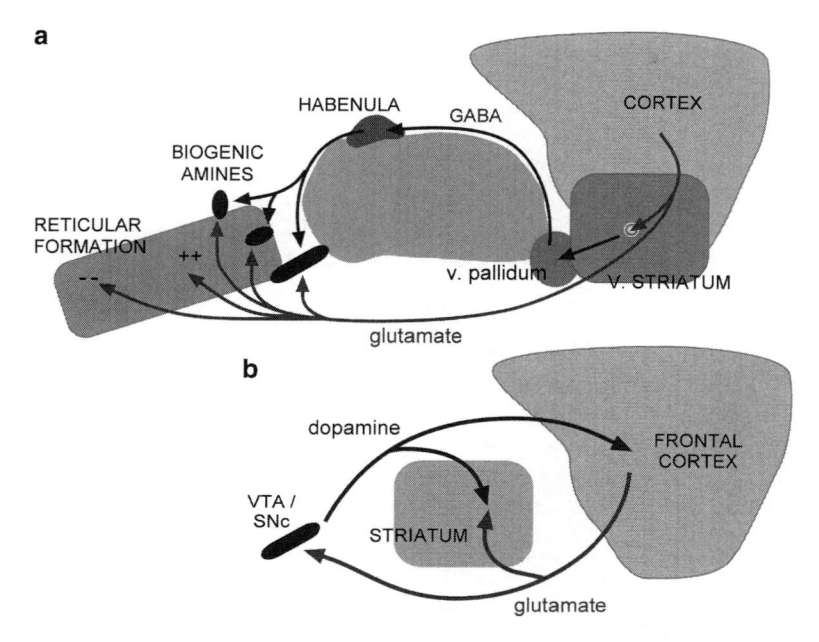

Fig. 1 Cortical control of reticular formation and biogenic amine systems. (**a**) Two potential pathways over which frontal cortex can exert a potent influence on biogenic amine nuclei (*Dark ellipsoids*: substantia nigra, pars compacta; raphe; locus coeruleus) are illustrated. Descending glutamatergic projections from cortex innervate the nuclei directly and indirectly via a complex circuit that begins with the ventral striatum. The latter structure has massive connections with the ventral pallidum, which sends a surprisingly large projection via the stria medullaris to the habenula. The habenula in turn has extensive efferents to the biogenic amine nuclei. Note the prevalence of long, GABAergic axons in the indirect pathway. Long descending projections from frontal cortex also reach the excitatory (++) and inhibitory subdivisions (− −) of the reticular formation. There is a wealth of physiological data indicating that the illustrated descending pathways result in a strong cortical influence over the brainstem regions. (**b**) One of the multiple "loops" formed between the brainstem regions indicated in panel **a** and frontal cortex. Dopaminergic neurons in the ventral tegmental area and substantia nigra pars compacta (VTA/SNc) send ascending projections to striatum and frontal cortex. The latter area then projects back to VTA/SNc as well as to striatum to a degree that, in the hypothesis, reflects the intensity of the incoming dopaminergic signal. A second type of feedback depends upon circuitey within panel **b**. In this, VTA/SNc projections to striatum modify activity in that structure which results in altered output in the striatal > dorso-medial nucleus (thalamus) > frontal cortex circuit (not illustrated). The frontal cortex then sends a signal to the striatum that is adjusted according to the intensity of alterations within striatum produced by the ascending dopaminergic system. The neocortex thus, according to the hypothesis, counteracts dopamine-driven shifts in striatal functions

biochemical pathways associated with spine morphology, and thus the spine cytoskeleton, are disrupted by any of a long list of conditions. In accord with this idea, several of the more than 40 gene products linked to mental retardation interact directly, as regulatory factors or downstream effectors, with the Rho family of small GTPases [52, 53]. Still others appear to be related to various links between Rho GTPases and the cytoskeleton [54, 55]. The Rho family regulates actin filament assembly and has potent effects on spine morphology during development [56]. Several investigators have concluded from these results that disturbances to actin signaling may be involved in diverse conditions involving memory and cognitive impairments.

The advent of light microscopic techniques for visualizing both increases in filamentous (F-) actin and protein state (e.g., conformation or phosphorylation) at individual synapses in appropriately prepared brain sections resulted in the first pictures of activity-induced actin remodeling in adult spines [13, 26, 57, 58]. With infusion of fluorescent-tagged phalloidin into the living hippocampal slice during or soon after the induction of LTP, activity-dependent increases in F-actin were visualized in a subset of spines in the field of afferent stimulation [57] (Fig. 2). Other studies using phospho-specific antisera to receptors and signaling proteins demonstrated that LTP-inducing stimulation activates actin regulatory elements in this same subset of spines [59, 60]. As seen from the latter immunocytochemical studies, the synapse has two classes of receptors in addition to those mediating transmission, one group that provides for adhesion and a second group of

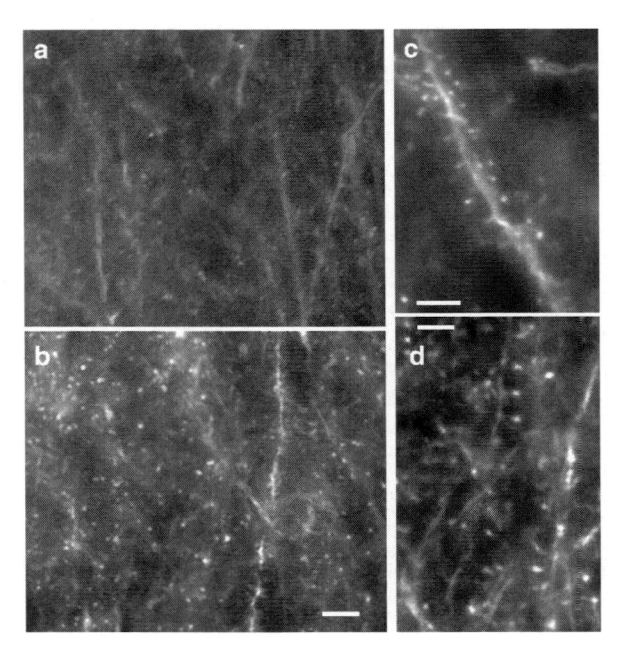

Fig. 2 Learning-related patterns of afferent activity cause actin polymerization within dendritic spines in adult hippocampus. Phalloidin, a compound that labels filamentous (F-), but not monomeric, actin was topically applied to the dendritic zone of interest at the conclusion of physiological testing prior to fixation ("in situ labeling"). (**a**) Labeling in hippocampus from a slice that received low frequency stimulation throughout the testing period. (**b–d**) A single train of ten theta bursts caused a marked increase in the number of dense, F-actin positive puncta (bar = 10 μm for **a** and **b**). (**c**) The size and distribution of the puncta along fine dendritic branches indicates that they are spines (bar = 10 μm). (**d**) Thin necks connecting the spine heads to the dendrite are evident in some cases (bar = 5 μm). Modified from Lynch et al. [26]

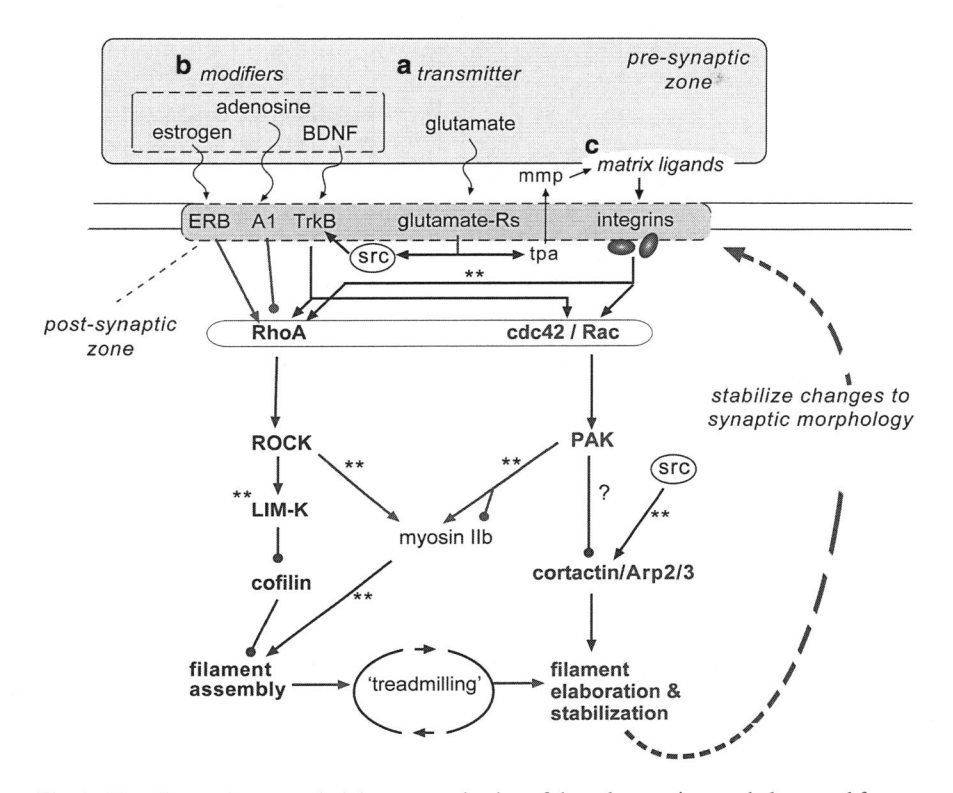

Fig. 3 Signaling pathways underlying reorganization of the subsynaptic cytoskeleton and factors that regulate them. The postsynaptic specialization in this model contains transmembrane receptors for three classes of extracellular signals: (**a**) the neurotransmitter; (**b**) releasable "modifiers"; (**c**) molecules that generate the strong adhesive property of synapses. This last group is denoted by integrins because of recent work describing the pivotal role of these receptors in regulating actin dynamics at glutamatergic synapses in the adult hippocampus [117] (other adhesion receptors such as cadherins and NCAMs are likely involved as well). The ligands for the receptors arise from classical transmitter release [(**a**) glutamate], three presynaptic, releasable modifier factors [(**b**) estrogen, adenosine, and BDNF], and the extracellular matrix [(**c**) matrix ligands]. Theta pattern activity arriving at the presynaptic zone causes enhanced release of glutamate and secretion of all three modifiers. Activation of the postsynaptic glutamate receptors (AMPA- and NMDA-type) leads to dendritic release of activating agents ("tpa") for matrix metalloproteinases ("mmp") that create integrin ligands [118]. The glutamate receptors also engage Src family kinases, which then phosphorylate BDNF's TrkB receptor; the latter event combined with released BDNF results in full activation of the neurotrophin receptors [59]. Events following stimulation of the modifier and adhesion receptors are described in the lower part of the schematic. Signaling from both receptor classes converge on the Rho GTPases. The estrogen receptor beta (ERB) activates, whereas the adenosine receptor A1 inhibits, RhoA; neither receptor appears to significantly affect the related GTPases cdc42 and Rac. However, the third modifier, BDNF, activates all three GTPases. The GTPases initiate two actin signaling cascades, the first of which (RhoA > ROCK > LIMK) phosphorylates and thereby inactivates the actin severing protein cofilin; this event opens the way to actin filament assembly. Experimental work has confirmed that the RhoA path is critical to theta-driven actin polymerization in dendritic spines [60]. The newly formed polymers remain in a dynamic state in which they add monomers at one end and lose them at the other ("treadmilling") for several minutes until stabilized by the second of the GTPase-initiated signaling cascades (cdc/Rac > PAK > cortactin/Arp2/3). The last step, linking PAK to a protein

"modifiers" that strongly affects actin-related signaling (Fig. 3). While essentially all afferent activity affects probabilistic release, and thus probabilistic neurotransmitter receptor binding, only a narrow range of naturally occurring patterns of afferent input meaningfully engages the other two categories of receptors [59]. Available evidence indicates that activation of the modifier or adhesion receptors initiates two Rho GTPase-dependent signaling pathways that serve to trigger actin filament assembly (polymerization) and then to stabilize the newly formed polymers [60, 61]. These events are critical for the production of lasting changes of the subsynaptic cytoskeleton and memory-related changes to synaptic strength [13].

Two aspects of the model illustrated in Fig. 3 are pertinent to the possible use of ampakines in treating psychiatric conditions involving the spine cytoskeleton. First, acutely enhancing AMPA receptor-gated fast EPSCs increase the likelihood that patterned input will activate latent adhesion and modifier receptors and thereby set in motion signaling cascades leading to reorganization of spine cytoskeleton. Second, as noted, the drugs cause the up-regulation of BDNF [62, 63], one of the more potent of the synaptic modifiers; it is reasonable to assume that this will also result in greater than normal actin signaling within spines [58]. Hypothetically, these effects could offset the debilitating effects of diseases on cytoskeletal dynamics.

To summarize, there are reasons to assume that acute treatments with ampakines will reduce symptoms associated with biogenic amine-based disorders while chronic or spaced administration will have positive effects in conditions involving spine cytoskeletal abnormalities. In the latter case, the beneficial effects of treatment would not require continuous presence of the drug; i.e., increased levels of endogenous BDNF will outlast their initiating conditions. The following sections survey results from various tests of these arguments.

3 Ampakines and Abnormal Biogenic Amine Activity

3.1 Stimulant-Induced Hyperactivity

The first attempt to counteract disturbances arising from aberrant biogenic amine activity examined the behavioral stereotypy produced in rats by methamphetamine

---

Fig. 3 (continued) complex that initiates filament branching and stabilization, has not been studied at synapses. However, activation of the cortactin/Arp2/3 complex is driven by Src, a kinase group that is activated at synapses during LTP induction [59]. Thus, various lines of evidence indicate that the cdc42/Rac pathway plays a central part in stabilizing the actin filaments formed after theta stimulation. Together, these protein sequences and actin management steps result in a new cytoskeleton that serves to anchor the observed changes in synaptic morphology, and thus synaptic potency, which occur shortly after the induction stage of LTP. Experimental evidence (immunofluorescence labeling at synapses, pharmacology, biochemistry) is available for most of the elements and sequences illustrated in the diagram and, for the large part, are described in publications [57, 59, 60, 119]. Other features (*double asterisk*) are well established for nonneuronal cells but have not been directly studied at adult synapses

injections. An early ampakine with modest potency and a short half-life reduced stereotyped rearing by about 70%, with the effect wearing off as the compound was eliminated [64]. A subsequent study confirmed that ampakines reduce the hyperactivity produced by moderate doses of methamphetamine and further showed that the potency of the compounds is highly correlated (r^2 of ~0.80) with their enhancing effects on AMPAR-gated currents [65]. Moreover, ampakines proved to be synergistic with antipsychotic drugs, in that a combination of subthreshold concentrations of both types of compound reduced locomotor scores after amphetamine injections almost to levels found in vehicle-treated rats [65]. This is as expected from the cortical feedback models: partial blockade of dopamine receptors (antipsychotics) should positively interact with a modest increase in glutamatergic transmission. Ampakines did not, however, enhance the catalepsy produced by certain antipsychotic agents, indicating that they did not act on the primary dopaminergic effects of the latter compounds.

The above results cannot be attributed to a generalized depressive effect of ampakines on rat locomotor activity. Dosages that do not measurably change baseline exploratory activity are nonetheless effective in amplifying the antistimulant properties of antipsychotic drugs [65]. More generally, the influence of ampakines on behavioral activity is situation-dependent. The drugs reduce the frequency of movement in rats investigating a new environment [18] as anticipated from the assumption that they modulate the high level of arousal engendered by such circumstances. But the same compounds, at the same doses, actually *increase* the speed of movement and the frequency of searching responses when rats are dealing with a well-learned maze in which variable reward sites must be identified [66]. Analyses of these types suggest that ampakines increase the cortical component of behavior, counteracting the impulses produced by lower brain systems while emphasizing the shaping effects of frontal and hippocampal regions.

3.2 Schizophrenia and Attention Deficit/Hyperactivity Disorder (ADHD)

Subsequent work tested for ampakine effects in rodent models of two psychiatric disorders generally assumed to involve hyperdopaminergic activity. Multiple versions of the drugs caused a dose-dependent, dramatic reduction of hyperactivity in a mouse model of ADHD [67]. Studies from another group found that ampakines counteract two diagnostic features of schizophrenia in a mouse model. The authors used a conditioned emotional response paradigm in which the mutants had major deficits in latent inhibition and prepulse inhibition, measures that are commonly abnormal in schizophrenics. Acute treatment with the ampakine CX546 returned both measures to levels found in wild-type mice [68]. Potent effects of ampakine drugs were also obtained in experiments using the Ungerstedt model of hyperdopaminergic activity [69, 70]. This paradigm involves a unilateral lesion of the εscending dopamine projections from the substantia nigra pars compacta: after a 2-week postlesion delay, systemic treatment with methamphetamine causes the animals to compulsively rotate in circles. The study showed that with acute treatment, two different ampakines (CX546, CX614) reduced this compulsive turning behavior at concentrations predicted to be effective from their effects on glutamatergic transmission in brain slices [30]. In two phencyclidine (PCP) models of schizophrenia, acute administration of the ampakine CX516 reduced cognitive deficits in a test examining "set shifting" (i.e., the ability to switch cognitive strategies in response to changes in the environment) [71].

Positive results have also been reported for a phase II clinical trial involving adult ADHD patients. Using a multidose, cross-over design, subjects received both an ampakine (CX717) and placebo for 3 weeks with a washout period interposed between the treatments. The AMPAR modulator, at the predicted dosage, caused a sizeable reduction in scores on both the attention deficit and hyperactivity components of a standard rating scale.

3.3   Depression

Much less work on the possibility that enhancing glutamatergic transmission in cortex will normalize abnormal activity in ascending biogenic amine projections has been done for nondopaminergic systems. However, the few reports have been encouraging. Skolnick and colleagues found that AMPAR potentiators are effective in serotonin-dependent, learned helplessness models of depression [72–74]. Another group found that the compounds are similarly potent in a different test for depression in rat [75]. And again, as predicted by the feedback hypothesis, ampakines had a strong, synergistic interaction with antidepressant, serotonin-targeted drugs.

3.4   Cortical Effects of Ampakines

Feedback models predict that ampakines suppress dopamine hyperactivity via pronounced activation of cortex. The first investigation of this idea compared the effects of methamphetamine with those of an ampakine on expression of the activity-regulated gene c-fos in neocortex and striatum [76]. The stimulant produced marked increases in c-fos mRNA within dorsal and ventral striatum with smaller effects in orbitofrontal cortex; there were no measurable changes in anterior (dorsal) cortex. In contrast, the ampakine enhanced c-fos expression in dorsal neocortex, caused little if any change in orbitofrontal areas, and modestly reduced c-fos mRNA levels in striatum. Collectively, the data indicate that ampakines shift the ratio of cortico-striatal activity in favor of the cortex while methamphetamine has the opposite effect [76].

Brain activity data were also assessed as part of the studies using the Ungerstedt paradigm. In methamphetamine-treated rats, cotreatment with ampakines increased *c-fos* expression in the forelimb area of cortex, but not in striatum [30]. The ampakine effects were regionally selective across cortical subfields and larger ipsilateral to the nigro-striatal lesion leading to a greater symmetry of cortical activation in 6-OH-DA lesioned rats. Thus, the drugs had their greatest effects on aggregate activity in cortical regions needed to suppress the compulsive turning behavior produced by hyperdopaminergic activity.

While activity mapping results were not collected in studies of biogenic amine disorders, it is of interest with regard to feedback hypotheses that ampakine-induced increases in cortical activity have been observed in primates [77]. The pertinent experiments involved PET scans of rhesus monkeys performing a complex, but well-learned, delayed-match-to-sample problem. The task resulted in intense metabolic activity in dorso-lateral frontal cortex and temporal lobes under control conditions; the ampakine substantially increased activity in both cortical regions; moreover, the precuneus was engaged during testing in the presence of the drug. The effects on brain activity were associated with behavioral scores well above those achieved by the monkeys under placebo conditions. These findings confirm that observations in rodents extend to primates and provide a particularly clear-cut example of how ampakines can act in a regionally selective manner.

4 Effects of Ampakines on Conditions in Which Spine Plasticity Is Impaired

While spine abnormalities are found in multiple instances of mental retardation, it is not clear that they are a common feature of disorders involving memory and cognition. This point prompted us to examine physiological and cytoskeletal plasticity in several models of human conditions associated with cognitive impairment. The most extensive research so far involves Huntington's disease (HD), a progressive neuropathological disease that has been traced to a single gene mutation [78] and for which multiple mouse analogs are available. But, as described below, work on other conditions involving memory problems also has been informative.

4.1 Huntington's Disease

Early stage HD is characterized by significant defects in memory and cognition and it has been argued that these impairments are among the very first signs of the disease [79, 80]. Consonant with this are reports that LTP is significantly impaired relatively early in the life of HD mouse models [81–83]. We reexamined this point using three transgenic HD mouse lines that do not develop pronounced motor problems before middle-age. Induction of LTP by the naturalistic theta burst stimulation (TBS) pattern was impaired in all three knock-in, transgenic strains [26, 27]. Detailed analyses showed that the plasticity deficit was selective in that baseline synaptic transmission could not be distinguished from that in wild-type animals. This also held for the composite EPSP responses to theta bursts and to the initial expression of LTP. It thus appears that the HD mutation acts at some stage of LTP beyond the extraordinarily complex events that go into the induction of the potentiation effect. From this it follows that the defect lies in those delayed signaling cascades that stabilize ("consolidate") LTP.

LTP consolidation requires reorganization of the spine cytoskeleton [13] and so we assessed TBS-induced actin filament assembly (polymerization) in HD mice. The results were clear: TBS caused a marked increase in spines containing high concentrations of F-actin in wild-type mice but not in the HD mutants (Fig. 4a) [26]. This constituted the first evidence that the Huntington mutation disrupts spine cytoskeletal plasticity. BDNF, which potently facilitates activity-driven actin polymerization, is decreased in HD patients and in mouse models of the disease [84]. If the simple loss of the neurotrophin or its availability to postsynaptic receptors is responsible for the loss of plasticity in HD mice, then brief infusions of BDNF should rescue both actin polymerization and LTP consolidation. Surprisingly, given the many effects associated with the Huntington mutation, we were able to confirm

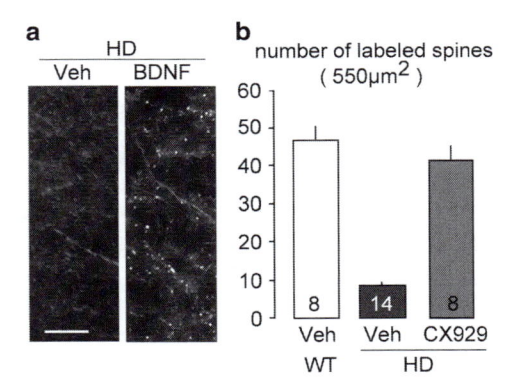

Fig. 4 Theta-driven actin polymerization in a mouse model of Huntington's disease (HD) is rescued by acute infusions of BDNF or chronic treatment with an ampakine. The "in situ" phalloidin method was used to label F-actin in spines in hippocampal slices prepared from HD model, CAG140 mice; a strain in which motor problems are relatively slight in early adulthood. (**a**) Theta bursts did not increase the number of densely labeled spines in vehicle (Veh)-treated CAG140 slices (bar = 10 μm). This result provided the initial evidence that actin signaling is disrupted by the HD mutation. A 60 min pretreatment with 2 nM BDNF restored activity-driven polymerization to levels seen in wild-type (WT) slices (see Fig. 2b for the response in wild-type slices; numbers on bar denote numbers of slices per group). (**b**) Four daily injections with a short half-life ampakine (CX929) returned TBS-induced actin polymerization in CAG140 mice to levels found in WT mice. Slices were prepared 18 h after the last CX929 injection. Densely labeled spines were counted using a blind, automated system [26, 27]

this prediction: a 60-min infusion of BDNF restored both TBS-induced actin polymerization and LTP consolidation in HD mice [26].

The above results set the stage for a test of whether ampakine-induced increases in BDNF production correct disease-related losses in spine plasticity. In doing these studies, we took advantage of the earlier discovery that brief treatments with ampakines can produce long-lasting elevations in BDNF protein levels [85, 86]. We therefore used daily injections of a short half-life ampakine ($\sim$15 min) to stimulate production of the neurotrophin and carried out testing on the day following the last injection. Under these conditions, in which the drug was long removed from the animal and BDNF was elevated, TBS-induced actin polymerization and LTP consolidation in the HD mutant hippocampus was fully restored to wild-type levels [27] (Fig. 4b).

Success in restoring spine plasticity necessarily raised the question of whether the ampakine treatments offset the memory problems found in early stage HD. We addressed this using an unsupervised learning paradigm in which mice learn a novel environment containing distant cues and within-field objects [87]. The experimental question was simply whether the animals would recognize the substitution of a local object. Wild-type mice spend a great deal of time investigating the new object but most HD mutants do not. However, ampakine-pretreated HD mice, with elevated BDNF protein concentrations, were indistinguishable from wild-type mice in this behavior [27]. The most straightforward interpretation of these results is that Huntington's is a disease in which the loss of BDNF's synaptic actions negatively impacts spine plasticity and thereby causes problems with mechanisms of information storage. Treatments that increase endogenous BDNF levels are thus logical therapeutics for the cognitive aspects of the disease. Ampakines are of particular interest in this regard because of the brevity of required treatment and safety profiles.

4.2 Aging

Despite intense interest in the question, very little is known about the causes of age-related losses in memory processing. Deterioration is by no means restricted to old age and indeed appears to progress throughout adult life [88]. The effects of age are likely to be general to mammals as significant losses are present in early middle age in rats [89] and monkeys [90] as well as in humans. There is, in other words, a real possibility that brain aging, as assessed by memory performance, begins during young adulthood and continues steadily thereafter in rat as in man.

LTP is a reasonable starting point for a search into the causes of age-dependent decline in memory but most of the relevant research in this area has focused on old age, and it is not clear that the findings necessarily apply to the progressive impairments emerging earlier in life. Accordingly, we tested LTP in rats at early middle-age (9–12 months) and obtained a set of surprising results [91, 92]. Potentiation in the apical dendrites of hippocampal field CA1, the zone most commonly examined in plasticity experiments, was not markedly different from that seen in young adults (2–3 month old). However, LTP in the basal dendrites of the same target neurons failed to stabilize [91] (Fig. 5a). Baseline synaptic responses and paired–pulse facilitation, a measure of transmitter release kinetics, were comparable in young and middle-aged animals, so the plasticity defect was not secondary to a broad deterioration of synaptic functioning. This observation pointed to the possibility that the age-related loss was due to a defect in the spine cytoskeletal machinery that underlies LTP consolidation.

While the idea that progressive memory loss might be thought of as a spine disorder has received little attention, there are reasons to think that this might be the case. Of particular interest is the growing evidence that aging negatively affects the regulation of extracellular adenosine [93], one of the releasable "synaptic modifiers" known to regulate and potentially disrupt spine cytoskeletal reorganization and LTP consolidation (see Fig. 3) [60]. Tests for adenosine involvement in middle-aged impairments in LTP proved positive: infusions of an antagonist for adenosine's A1 receptor immediately after LTP induction restored potentiation to the same size and stability found in young adults [91]. Further work showed that adenosine clearance from the extracellular space was, as suspected, slower in the basal dendritic field of middle-aged animals than in young adults [91]. As concluded for tests with HD model mice, plasticity losses during early aging appear to arise not so much from a breakdown in actin dynamics within spines but rather from abnormalities in levels of a releasable factor which influences activity-dependent spine actin remodeling. In both cases, the complex system governing activity-driven spine/synapse reorganization is defective.

The work described immediately above set the stage for further tests of the hypothesis that positive modulation of AMPA receptors can be used to offset symptoms arising from spine impairments. To repeat, there are two currently recognized routes over which the drugs could be useful: acutely increasing the contributions of NMDA receptors to cytoskeletal reorganization and chronically up-regulating concentrations of the positive modifier BDNF (see above). We tested both ideas. Using hippocampal slices, an acute 30-min pretreatment with an ampakine was found to fully restore LTP in the CA1 basal dendrites of middle-aged rats to levels found in young adult rats [91] (Fig 5b). There is much that remains to be done in this line of work but the results as they stand confirm the basic hypothesis that amplifying AMPAR-gated synaptic currents can be used to offset unrelated failures in the cytoskeletal machinery that stabilizes synaptic changes.

Tests of the second route – that involving BDNF – used the paradigm employed in the above-described HD studies. Middle-aged rats were treated with a short half-life ampakine (CX929) for 4 days to up-regulate endogenous BDNF production; physiological testing was carried out 1 day after the last injection, a time point at which BDNF levels were elevated by 60% and the ampakine was long removed from the brain [92]. Baseline synaptic physiology and initial expression of LTP were normal in the CX929-treated middle-aged rats, in line with the findings for HD mice. Nonetheless, the otherwise unstable LTP in the basal dendrites of the drug-treated middle-aged animals was restored to the stability and magnitude found in young adults [92] (Fig 5c). Combining these results with those obtained with

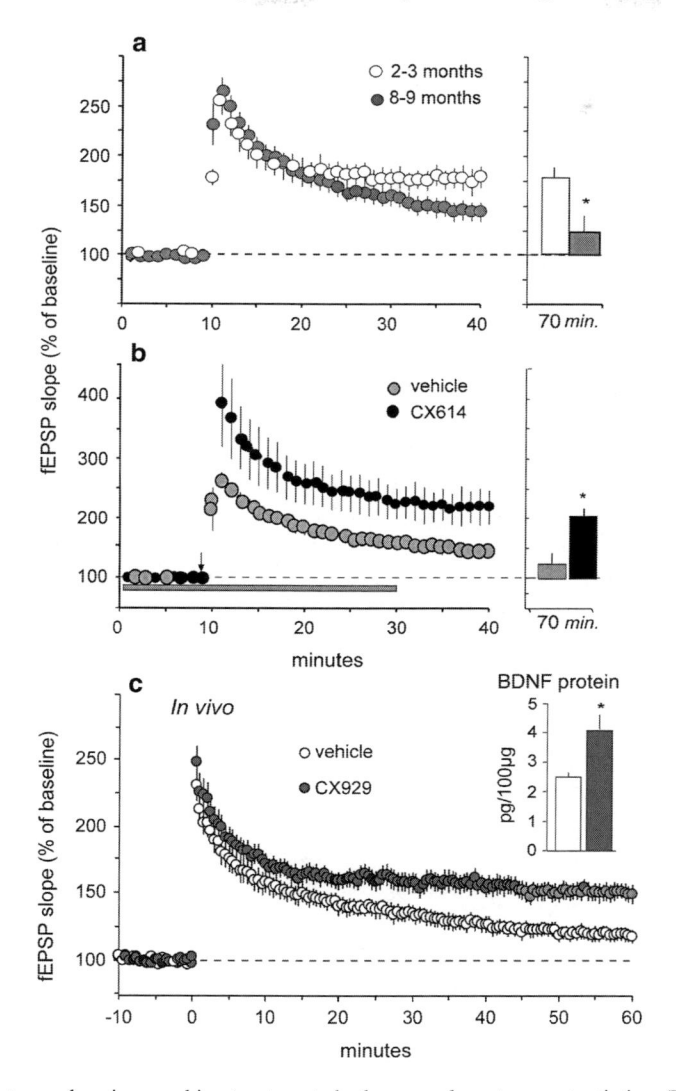

Fig. 5 Acute or chronic ampakine treatments both rescue long-term potentiation (LTP) in the hippocampus of middle-aged rats. Synaptic responses (field EPSPS) were collected from the basal dendrites of hippocampal field CA1 in slices prepared from middle-aged (8–9 months) or young adult (2–3 months) rats. (**a**) A single theta burst stimulation (TBS) train produced robust LTP in the basal dendrites of young, adult slices but resulted in steadily decaying potentiation in middle-aged slices. EPSPS at 70 min post-TBS were well above baseline in the younger group but had returned almost to baseline in the older cases (*bar graph at right*). Note in the left hand panel that the initial expression of potentiation is not detectably different in the two age groups for the first 2–4 min post-TBS; aging thus appears to act on events that stabilize LTP. (**b**) Acute treatment of middle-aged slices with an ampakine (CX614) rescues LTP in middle-aged slices. Percent LTP at 70 min post-TBS (*right* side) was equal to that found in young adults (see panel **a**). (**c**) Daily injections of the short half-life ampakine CX929 elevated BDNF levels and normalized LTP in middle-aged rats. A 4-day regimen of twice daily injections of vehicle or ampakine was used with slices

HD mice suggest that ampakine-induced increases in BDNF protein levels can overcome problems arising from an imbalance of positive (BDNF in HD) or negative (adenosine in aging) factors that influence synaptic signaling cascades controlling the spine actin cytoskeleton.

4.3 Chronic Reductions in Circulating Estrogen

The results for HD and aging constitute evidence that spine defect disorders, including those reflecting factors that regulate spine cytoskeletal reorganization, are more common than previously thought. We expanded this argument with an analysis of effects of chronic reductions in estrogen, another condition associated with memory problems. Numerous studies have found that complaints involving memory are commonplace after surgical ovariectomy [94, 95]; moreover, while the issue is controversial, menopause sometimes results in similar, clinically meaningful, difficulties [96, 97]. There are no widely accepted explanations for these effects but estrogen is known to have potent effects on submembrane actin networks in non-neuronal cells. It thus seemed possible that the hormone acts as a positive modifier of the LTP-related cytoskeletal machinery described in Fig. 3. If this were the case, then the loss of estrogen either through surgery or aging could create a spine plasticity disorder and thus the possibility of treatment with positive modifiers.

In accord with these ideas, we were able to completely block estrogen's well-established facilitatory effects on EPSPs by pretreating the hippocampus with a toxin that prevents actin monomers from assembling into filaments [61]. Follow-on studies provided the first evidence that estrogen causes a rapid but fully reversible (upon washout) actin polymerization response within dendritic spines and does so via activation of a cascade involving the small GTPase RhoA, its effector ROCK, and the actin regulatory protein cofilin [61]. However, while estrogen engages one of the pathways set in motion by the induction of LTP (see Fig. 3), it had little if any effect on a second cascade involving the small GTPases Rac and Cdc42, and their effector p21-activated kinase (PAK) [61]. Together, these results indicate that estrogen triggers a weak version of the cytoskeletal changes associated with LTP, and thus enhances fast EPSPs, but fails to initiate those events that stabilize newly formed actin networks and synaptic potentiation. Thus, the acute effects of estrogen are reversible.

The various discoveries converging on the conclusion that estrogen utilizes aspects of the signaling and actin machinery underlying LTP reinforce the idea that clinical conditions resulting in subnormal levels of the steroid create functional defects in spine cytoskeletal plasticity. Tests of this proved positive: ovariectomy in

---

Fig. 5 (continued) prepared 1 day after the last injection. Hippocampal BDNF levels were ~60% greater in the ampakine group than in the vehicle controls (inset *graph*). Theta stimulation produced decremental potentiation in the vehicle group but robust and stable LTP in the ampakine cases [91, 92]

rats depressed the RhoA pathway with the expected losses of activity-driven actin polymerization and LTP stabilization [61]. Remarkably, given that weeks of low estrogen levels were involved in these experiments, a brief infusion of the hormone fully rescued cytoskeletal and physiological plasticity [61]. It should be noted that hippocampal neurons synthesize estrogen [98] and that effects of ovariectomy on neuroplasticity could be secondary to a depression of local hormone production. In any case, the experiments point to the conclusion that estrogen losses result in a spine defect disorder.

The question then became one of whether the broad strategy of using positive modifiers, and in particular BDNF, to normalize spine cytoskeletal reorganization is applicable in the particular case of low estrogen. We investigated this with the same experimental sequence used for in the early stage HD and aging studies. Acute infusions of low concentrations of BDNF rescued TBS-driven actin polymerization in pyramidal cell dendritic spines within hippocampal slices from ovariectomized (OVX) rats, and this effect was accompanied by a full recovery of LTP. Comparable restoration of actin remodeling and LTP was obtained in slices from OVX rats prepared 18 h after the last of four daily injections with an ampakine, a time point at which BDNF levels were elevated above control levels and the drug had left the system [99]. These results constitute what may be the clearest indication thus far that positive modifiers can substitute for each other in promoting the cytoskeletal changes underlying stable LTP. They also reinforce the feasibility of a broad-spectrum BDNF strategy based on minimally invasive ampakine treatments.

4.4 Fragile-X Mental Retardation Syndrome (FXS)

Unlike the previous cases, FXS has long been associated with spine abnormalities and, in particular, an excess of long, thin spines with a reduction in spines with bulbous heads [100, 101]. Qualitatively similar abnormalities were first described for cases of nonsyndromic mental retardation [102] and were attributed to a failure of spine maturation, an interpretation that is still common today. Morphologically aberrant spines have also been reported for various gene mutations, in addition to FXS, that are predisposing for retardation [49–51, 103]. It is of interest that some of these involve the Rho GTPases or their various effectors.

The Fragile-X mutation results in inactive Fragile X Mental Retardation Protein (FMRP) that regulates RNA translation and transport [104]. There is also evidence that FMRP interacts with certain protein transport systems [105–108]. Understanding how these effects relate to the FXS phenotype is the great challenge for research on the disorder. Aberrant spines are indicative of cytoskeletal disturbances and we tested for these in FMRP knockout mice. LTP was impaired in hippocampus [109], in accord with work in cortex [110–114], and appeared to be normalized by brief infusions of BDNF [109]. Surprisingly, given these results, activation of the RhoA–cofilin–actin polymerization cascade appeared normal with theta stimulation in the FMRP knockout hippocampus [109], a result suggesting that the essential defect in the disorder lies in the second signaling cascade mobilized by LTP induction (see Fig. 3). Recent work from our laboratory has confirmed this prediction [120]. It remains now to be determined if BDNF infusions correct the aberrant spine anatomy that characterizes FXS, and if chronic, ampakine-induced increases in BDNF content produce the same effects as the infusions of the neurotrophin.

5 Summary and Discussion

The invention of the ampakines was prompted by the assumption that facilitation of fast EPSCs in the cortical telencephalon would enhance memory and cognition by (1) promoting communication within neural networks underlying complex behavior and (2) lowering the threshold for induction of LTP. Subsequent work from different laboratories using a wide array of experimental arrangements has confirmed these predictions, at least for rodents and primates. The few studies involving normal human subjects, while very limited in scope, have also been encouraging. Side-effect profiles of the compounds across the animal and human experiments are usually reported to be benign. Accordingly, positive modulation of AMPA receptors has emerged as a plausible strategy for treating memory losses associated with aging and age-related neurodegenerative diseases.

The idea of using the compounds to treat neuropsychiatric disorders grew out of two very different literatures, one that dates back to fin de siecle Europe and the other involving contemporary work on neurotrophism. As noted, Hughlings-Jackson and his contemporaries reached the conclusion from clinical studies that the neocortex suppresses lower brain regions, an idea that gradually evolved into theories of feedback loops involving arousal/emotional systems and anterior neocortex. It follows from these concepts that either hyperactivity in the lower brain segments of the loops, or hypoactivity in the cortical components, will result in abnormal psychological states. A literal reading of the feedback loop theories also leads to the prediction that enhancing communication within cortex, an effect expected from ampakines, should suppress such abnormalities.

Links between neurotrophic signaling and ampakines were suggested by the discovery that intense neuronal activity causes a marked increase in the neuronal production of BDNF, a primary forebrain growth factor. Since the intended effect of positive modulators of AMPA receptors is to increase excitatory drive on target neurons, it follows that the drugs should up-regulate BDNF and related neurotrophins. Experimental work confirmed this prediction and, in addition, led to the discovery that brief periods of enhanced transmission can produce unexpectedly long-lasting changes in BDNF protein levels. The relevance of this to psychiatric disorders became evident when it was found that BDNF exerts potent effects over the subsynaptic cytoskeleton in the adult brain. A steadily growing body of results suggests that aberrant spine morphology and thus disturbances in the actin networks controlling spine anatomy are present in disparate neuropsychiatric disorders involving cognition. This relationship was first recognized for mental retardation but, as reviewed here, may be a factor in a broad array of clinical conditions. Ampakine-induced increases in expression and levels of the potent cytoskeletal modifier BDNF thus became a possible chronic treatment for diseases involving structural abnormalities within spines.

Experimental work confirmed that ampakines enhance cortical activity in rats and monkeys and, in agreement with Jacksonian ideas about hierarchical control in brain, found that this enhancement was associated with greater control over arousal, suppression of dopamine-driven compulsive turning, and performance on a complex behavioral problem. It is hypothesized that the elevated neuronal activity reflects recruitment of subthreshold neurons into the networks assembled to deal with the particular circumstances facing the animal. In support of this idea, minimal concentrations of ampakines increase throughput in complex cortical networks to a greater degree than would be expected from the monosynaptic actions of the compounds [16]. Moreover, behaviorally relevant dosages of ampakines elevate the activity of hippocampal neurons that under normal conditions are only marginally engaged by particular components of a difficult behavioral problem. Perhaps, the most extreme example of network expansion was obtained in monkey experiments in which ampakines allowed neocortex to engage an additional, functionally pertinent region during performance of a cognitively demanding task [77].

A striking feature of the work on cortical activation is that ampakines act in a regionally selective manner that reflects the task demands placed on the subject. They enhanced aggregate activity in frontal cortex during exploration, in somatosensory cortex during compulsive turning, and in frontal and temporal neocortex in animals working at the limits of their cognitive abilities. It is not unreasonable to conclude from these observations that the drugs, acting as receptor modulators, exert their greatest effects in "now-active" networks. It should be added that Arai and colleagues have found that ampakines can have markedly different potencies across brain regions (e.g., thalamus vs. cortex) [10], most likely because of preferences for particular AMPA receptor combinations. It remains to be determined how large a role this type of selectivity plays in cortex where the most prominent receptor differences appear to be laminar in nature and to involve splice variants of receptor subunits.

Clearly missing from the animal model work on ampakines and biogenic amine-related disorders are unit recording data on the degree to which the drugs affect activity in the brainstem raphe, locus coeruleus, and ventral tegmental area/substantia nigra. It is possible that cortical activation modulates activity in these areas via pathways of the type shown in Fig. 1 or, alternatively, offsets the effects of the ascending systems within their upstream targets. Mapping of ampakine-induced changes in aggregate neuronal activity across the entire brain would also be of utility in addressing questions about potential sites of interaction. Attempts to evaluate arguments about ampakines and biogenic amine-related disorders are also seriously constrained by a lack of clinical results. Published work on schizophrenia used a drug variant with modest potency and a short half-life, features that complicate the evaluation of the generally negative results. A more potent, longer lasting variant was reported to produce substantial, rapidly developing improvements in adult ADHD patients. But much more work, possibly with still more potent compounds, will be needed to reach a first analysis of the broad-spectrum treatment hypothesis.

The BDNF component of the ampakine strategy also lacks critical data. For one, it has yet to be demonstrated that up-regulating the neurotrophin concentrations in vivo actually enhances BDNF signaling across synapses. However, development of the dual immunofluorescence microscopy techniques used to analyze signaling at very large numbers (millions) of individual synapses (see above) has opened the way to direct tests of the question. Work of this kind has already shown that with ampakine-induced increases in BDNF expression in cultured hippocampal slices there is a significant increase in activation (phosphorylation) of BDNF's TrkB receptor at spine synapses [86]. Further studies have shown that in adult rat hippocampus LTP induction and learning of complex environments, both trigger TrkB activation in a BDNF-dependent fashion [59]. Moreover, the latter study demonstrated that theta rhythm activity, as occurs during learning, is particularly effective for activating synaptic TrkB. It should now be possible to use the same analytical techniques test if elevating BDNF concentrations in vivo lowers the threshold and/or increases the magnitude of TrkB signaling elicited by LTP or learning.

Beyond this lies the question of how much of the BDNF effect in the ampakine experiments is due to the neurotrophin itself as opposed to factors it induces. In favor of a direct effect is evidence that acute BDNF infusions rescue actin poly-merization and LTP consolidation in spine defect models in a manner comparable to that found after up-regulation by daily ampakine treatment. Still, BDNF is known to influence the expression of other substances, such as VGF [115] and neuropep-tide Y [116], with known neuroprotective or neuromodulatory actions; the question of direct and indirect actions will likely assume increasing importance as more is learned about how ampakine-induced elevations affect neuropathological condi-tions. Finally, it need not be stressed that there are no readily available clinical data on the effects of ampakines on BDNF levels in humans or on the consequences of several days of treatment for disorders involving spine defects. Extrapolating from current clinical findings with regard to safety, possible BDNF-like effects, etc. needs to be done with caution as there are marked differences between ampakines with regard to the extent that they enhance BDNF production. This appears to be related to the degree to which the drugs prolong the duration of EPSCs, and thus on the degree to which they slow receptor desensitization as well as deactivation.

In all, ampakines have become reasonably well-characterized (binding pocket, receptor kinetic effects, physiology, etc.) drugs whose observed functional effects can be accounted for, without much violence to the data, by a small set of hypotheses. These same ideas predict that the compounds will have positive actions in a surprisingly large number of neuropsychiatric disturbances, and this has been borne out by a limited amount of preclinical work. But these same studies have raised new and important issues and, as indicated, clinical data are rare. A clear idea of the drugs' potential will likely evolve over the next several years.

Acknowledgments Research described in this commentary was supported by National Institutes of Neurological Disorders and Stoke grants NS45260 and NS051823 to G.L. and C.M.G. and National Institutes of Mental Health grant MH083346 to C.M.G. and J.C.L.

References

1. Nowak L, Bregestovski P, Ascher P, Herbet A, Prochiantz A (1984) Magnesium gates glutamate-activated channels in mouse central neurones. Nature 307:462–465
2. Nicoll RA (2003) Expression mechanisms underlying long-term potentiation: a postsynaptic view. Philos Trans R Soc Lond B Biol Sci 358:721–726
3. Lynch G (1998) Memory and the brain: unexpected chemistries and a new pharmacology. Neurobiol Learn Mem 70:82–100
4. Abraham WC, Williams JM (2003) Properties and mechanisms of LTP maintenance. Neuroscientist 9:463–474
5. Gall CM, Lauterborn JC (2000) Regulation of BDNF expression: multifaceted, region-specific control of a neuronal survival factor in the adult CNS. In: Mocchetti I (ed) Neurobiology of the neurotrophins. FP Graham Publishing Co., Johnson City, TN, pp 541–579
6. Castren E, Berninger B, Leingartner A, Lindholm D (1998) Regulation of brain derived neurotrophic factor mRNA levels in hippocampus by neuronal activity. Prog Brain Res 117:57–64
7. Staubli U, Rogers G, Lynch G (1994) Facilitation of glutamate receptors enhances memory. Proc Natl Acad Sci USA 91:777–781
8. Staubli U, Perez Y, Xu F, Rogers G, Ingvar M, Stone-Elander S, Lynch G (1994) Centrally active modulators of glutamate (AMPA) receptors facilitate the induction of LTP in vivo. Proc Natl Acad Sci USA 91:11158–11162
9. Jin R, Clark S, Weeks AM, Dudman JT, Gouaux E, Partin KM (2005) Mechanism of positive allosteric modulators acting on AMPA receptors. J Neurosci 25:9027–9036
10. Arai AC, Kessler M (2007) Pharmacology of ampakine modulators: from AMPA receptors to synapses and behavior. Curr Drug Targets 8:583–602
11. Lynch G, Gall CM (2006) Ampakines and the threefold path to cognitive enhancement. Trends Neurosci 29:554–562
12. Lynch G (2006) Glutamate-based therapeutic approaches: ampakines. Curr Opin Pharmacol 6:82–88
13. Lynch G, Rex CS, Chen LY, Gall CM (2008) The substrates of memory: defects, treatments, and enhancement. Eur J Pharmacol 585:2–13
14. O'Neill MJ, Witkin JM (2007) AMPA receptor potentiators: application for depression and Parkinson's disease. Curr Drug Targets 8:603–620
15. Ryder JW, Falcone JF, Manro JR, Svensson KA, Merchant KM (2006) Pharmacological characterization of cGMP regulation by the biarylpropylsulfonamide class of positive, allosteric modulators of alpha-amino-3-hydroxy-5-methyl-4-isoxazolepropionic acid receptors. J Pharmacol Exp Ther 319:293–298
16. Sirvio J, Larson J, Quach CN, Rogers GA, Lynch G (1996) Effects of pharmacologically facilitating glutamatergic transmission in the trisynaptic intrahippocampal circuit. Neuroscience 74:1025–1035
17. Arai A, Kessler M, Rogers G, Lynch G (1996) Effects of a memory enhancing drug on AMPA receptor currents and synaptic transmission in hippocampus. J Pharmacol Exp Ther 278:627–638
18. Granger R, Staubli U, Davis M, Perez Y, Nilsson L, Rogers GA, Lynch G (1993) A drug that facilitates glutamatergic transmission reduces exploratory activity and improves performance in a learning-dependent task. Synapse 15:326–329

19. Larson J, Lieu T, Petchpradub V, LeDuc B, Ngo H, Rogers GA, Lynch G (1995) Facilitation of olfactory learning by a modulator of AMPA receptors. J Neurosci 15:8023–8030
20. Rogan MT, Staubli UV, LeDoux JE (1997) AMPA receptor facilitation accelerates fear learning without altering the level of conditioned fear acquired. J Neurosci 17:5928–5935
21. Ingvar M, Ambros-Ingerson J, Davis M, Granger R, Kessler M, Rogers GA, Schehr RS, Lynch G (1997) Enhancement by an ampakine of memory encoding in humans. Exp Neurol 146:553–559
22. Ren J, Ding X, Funk GD, Greer JJ (2009) Ampakine CX717 protects against fentanyl-induced respiratory depression and lethal apnea in rats. Anesthesiology 110:1364–1370
23. Ren J, Poon BY, Tang Y, Funk GD, Greer JJ (2006) Ampakines alleviate respiratory depression in rats. Am J Respir Crit Care Med 174:1384–1391
24. Greer JJ, Ren J (2009) Ampakine therapy to counter fentanyl-induced respiratory depression. Respir Physiol Neurobiol 168:153–157
25. Ogier M, Wang H, Hong E, Wang Q, Greenberg ME, Katz DM (2007) Brain-derived neurotrophic factor expression and respiratory function improve after ampakine treatment in a mouse model of Rett syndrome. J Neurosci 27:10912–10917
26. Lynch G, Kramar EA, Rex CS, Jia Y, Chappas D, Gall CM, Simmors DA (2007) Brain-derived neurotrophic factor restores synaptic plasticity in a knock-in mouse model of Huntington's disease. J Neurosci 27:4424–4434
27. Simmons DA, Rex CS, Palmer L, Pandyarajan V, Fedulov V, Gall CM, Lynch G (2009) Up-regulating BDNF with an ampakine rescues synaptic plasticity and memory in Huntington's disease knockin mice. Proc Natl Acad Sci USA 106:4906–4911
28. Jourdi H, Hamo L, Oka T, Seegan A, Baudry M (2009) BDNF mediates the neuroprotective effects of positive AMPA receptor modulators against MPP(+)-induced toxicity in cultured hippocampal and mesencephalic slices. Neuropharmacology 56:876–885
29. O'Neill MJ, Murray TK, Whalley K, Ward MA, Hicks CA, Woodhouse S, Osborne DJ, Skolnick P (2004) Neurotrophic actions of the novel AMPA receptor potentiator, LY404187, in rodent models of Parkinson's disease. Eur J Pharmacol 486:163–174
30. Hess U, Whalen S, Sandoval L, Lynch G, Gall C (2003) Ampakines reduce methamphet-amine-driven rotation and activate neocortex in a regionally selective fashion. Neuroscience 121:509–521
31. York G, Steinberg D (2006) An introduction to the life and work of John Hughlings Jackson with a catalogue raisonné of his writings. Med Hist Suppl 26:3–157
32. Dell P (1963) Reticular homeostasis and critical reactivity. In: Moruzzi G, Fessard A, Jasper H (eds) Brain mechanisms. Elsevier, New York, pp 82–103
33. Bonvallet M, Hugelin A (1961) Influence de la formation reticulaire et du cortex cerebral sur l'excitabilite motrice au cours de Phypoxie. Electroencephalogr Clin Neurophysiol 13:270–284
34. Hugelin A, Bonvallet M, Dell P (1959) Activation reticulaire et corticale d'origine chemo-ceptive au cours Phypoxie. Electroencephalogr Clin Neurophysiol 11:325–340
35. Groves PM, Wilson CJ, Boyle RD (1974) Brain stem pathways, cortical modulation, and habituation of the acoustic startle response. Behav Biol 10:391–418
36. Fahn S (2008) The history of dopamine and levodopa in the treatment of Parkinson's disease. Mov Disord 23(Suppl 3):S497–S508
37. Anden NE, Carlsson A, Dahlstroem A, Fuxe K, Hillarp NA, Larsson K (1964) Demonstra-tion and mapping out of nigro-neostriatal dopamine neurons. Life Sci 3:523–530
38. Lynch G, Smith RL, Robertson R (1973) Direct projections from brainstem to telencephalon. Exp Brain Res 17:221–228
39. Steriade M (1996) Arousal: revisiting the reticular activating system. Science 272:225–226
40. Lynch GS, Ballantine P 2nd, Campbell BA (1969) Potentiation of behavioral arousal after cortical damage and subsequent recovery. Exp Neurol 23:195–206

41. Carlsson M, Carlsson A (1990) Interactions between glutamatergic and monoaminergic systems within the basal ganglia – implications for schizophrenia and Parkinson's disease. Trends Neurosci 13:272–276

42. Carlsson A, Hansson LO, Waters N, Carlsson ML (1999) A glutamatergic deficiency model of schizophrenia. Br J Psychiatry Suppl 37:2–6

43. Carlsson A (1995) Neurocircuitries and neurotransmitter interactions in schizophrenia. Int Clin Psychopharmacol 10(Suppl 3):21–28

44. Carlsson A (2006) The neurochemical circuitry of schizophrenia. Pharmacopsychiatry 39(Suppl 1):S10–S14

45. Purpura D (1975) Normal and aberrant neuronal development in the cerebral cortex of human fetus and young infant. UCLA Forum Med Sci 18:141–169

46. Irwin S, Patel B, Idupulapati M, Harris J, Crisostomo R, Larsen B, Kooy F, Willems P, Cras P, Kozlowski P et al (2001) Abnormal dendritic spine characteristics in the temporal and visual cortices of patients with fragile-X syndrome: a quantitative examination. Am J Med Genet 98:161–167

47. Rudelli R, Brown W, Wisniewski K, Jenkins E, Laure-Kamionowska M, Connell F, Wisniewski H (1985) Adult fragile X syndrome. Clinico-neuropathologic findings. Acta Neuropathol 67:289–295

48. Wisniewski K, Segan S, Miezejeski C, Sersen E, Rudelli R (1991) The Fra(X) syndrome: neurological, electrophysiological, and neuropathological abnormalities. Am J Med Genet 38:476–480

49. Marin-Padilla M (1976) Pyramidal cell abnormalities in the motor cortex of a child with Down's syndrome. A Golgi study. J Comp Neurol 167:63–81

50. Zhou Z, Hong E, Cohen S, Zhao W, Ho H, Schmidt L, Chen W, Lin Y, Savner E, Griffith E et al (2006) Brain-specific phosphorylation of MeCP2 regulates activity-dependent BDNF transcription, dendritic growth, and spine maturation. Neuron 52:255–269

51. Kaufmann WE, Moser HW (2000) Dendritic anomalies in disorders associated with mental retardation. Cereb Cortex 10:981–991

52. Chechlacz M, Gleeson JG (2003) Is mental retardation a defect of synapse structure and function? Pediatr Neurol 29:11–17

53. van Galen EJ, Ramakers GJ (2005) Rho proteins, mental retardation and the neurobiological basis of intelligence. Prog Brain Res 147:295–317

54. Ramakers G (2002) Rho proteins, mental retardation and the cellular basis of cognition. Trends Neurosci 25:191–199

55. Node-Langlois R, Muller D, Boda B (2006) Sequential implication of the mental retardation proteins ARHGEF6 and PAK3 in spine morphogenesis. J Cell Sci 119:4986–4993

56. Govek E, Newey S, Van Aelst L (2005) The role of the Rho GTPases in neuronal development. Genes Dev 19:1–49

57. Chen LY, Rex CS, Casale MS, Gall CM, Lynch G (2007) Changes in synaptic morphology accompany actin signaling during LTP. J Neurosci 27:5363–5372

58. Rex CS, Lin CY, Kramar EA, Chen LY, Gall CM, Lynch G (2007) Brain-derived neurotrophic factor promotes long-term potentiation-related cytoskeletal changes in adult hippocampus. J Neurosci 27:3017–3029

59. Chen LY, Rex CS, Sanaiha Y, Lynch G, Gall CM (2010) Learning induces neurotrophin signaling at hippocampal synapses. Proc Natl Acad Sci USA 107(15):7030–7035

60. Rex CS, Chen LY, Sharma A, Liu J, Babayan AH, Gall CM, Lynch G (2009) Different Rho GTPase-dependent signaling pathways initiate sequential steps in the consolidation of long-term potentiation. J Cell Biol 186:85–97

61. Kramár EA, Chen LY, Brandon NJ, Rex CS, Liu F, Gall CM, Lynch G (2009) Cytoskeletal changes underlie estrogen's acute effects on synaptic transmission and plasticity. J Neurosci 29:12982–12993

62. Lauterborn JC, Lynch G, Vanderklish P, Arai A, Gall CM (2000) Positive modulation of AMPA receptors increases neurotrophin expression by hippocampal and cortical neurons. J Neurosci 20:8–21

63. Legutko B, Li X, Skolnick P (2001) Regulation of BDNF expression in primary neuron culture by LY392098, a novel AMPA receptor potentiator. Neuropharmacology 40:1019–1027

64. Larson J, Quach CN, LeDuc B, Nguyen A, Rogers GA, Lynch G (1996) Effects of an AMPA receptor modulator on methamphetamine-induced hyperactivity in rats. Brain Res 738:353–356

65. Johnson S, Luu N, Herbst T, Knapp R, Lutz D, Arai A, Rogers G, Lynch G (1999) Synergistic interactions between ampakines and antipsychotic drugs. J Pharmacol Exp Ther 289:392–397

66. Davis CM, Moskovitz B, Nguyen MA, Arai A, Lynch G, Granger R (1997) A profile of the behavioral changes produced by facilitation of AMPA-type glutamate receptors. Psychopharmacology 133:161–167

67. Gainetdinov RR, Mohn AR, Bohn LM, Caron MG (2001) Glutamatergic modulation of hyperactivity in mice lacking dopamine transporter. Proc Natl Acad Sci USA 98:11047–11054

68. Lipina T, Weiss K, Roder J (2007) The ampakine CX546 restores the prepulse inhibition and latent inhibition deficits in mGluR5-deficient mice. Neuropsychopharmacology 32:745–756

69. Ungerstedt U (1976) 6-Hydroxydopamine-induced degeneration of the nigrostriatal dopamine pathway: the turning syndrome. Pharmacol Ther 2:37–40

70. Ungerstedt U (1971) Postsynaptic supersensitivity after 6-hydroxydopamine induced degeneration of the nigro-striatal dopamine system. Acta Physiol Scand 367:69–93

71. Broberg B, Glenthøj B, Dias R, Larsen D, Olsen C (2009) Reversal of cognitive deficits by an ampakine (CX516) and sertindole in two animal models of schizophrenia–sub-chronic and early postnatal PCP treatment in attentional set-shifting. Psychopharmacology (Berl) 206:631–640

72. Bai F, Li X, Clay M, Lindstrom T, Skolnick P (2001) Intra- and interstrain differences in models of "behavioral despair". Pharmacol Biochem Behav 70:187–192

73. Li X, Tizzano J, Griffey K, Clay M, Lindstrom T, Skolnick P (2001) Antidepressant-like actions of an AMPA receptor potentiator (LY392098). Neuropharmacology 40:1028–1033

74. Li X, Witkin J, Need A, Skolnick P (2003) Enhancement of antidepressant potency by a potentiator of AMPA receptors. Cell Mol Neurobiol 23:419–430

75. Knapp R, Goldenberg R, Shuck C, Cecil A, Watkins J, Miller C, Crites G, Malatynska E (2002) Antidepressant activity of memory-enhancing drugs in the reduction of submissive behavior model. Eur J Pharmacol 440:27–35

76. Palmer LC, Hess US, Larson J, Rogers GA, Gall CM, Lynch G (1997) Comparison of the effects of an ampakine with those of methamphetamine on aggregate neuronal activity in cortex versus striatum. Mol Brain Res 46:127–135

77. Porrino L, Daunais J, Rogers G, Hampson R, Deadwyler S (2005) Facilitation of task performance and removal of the effects of sleep deprivation by an ampakine (CX717) in nonhuman primates. PLoS Biol 3:e299

78. Vonsattel J, DiFiglia M (1998) Huntington disease. J Neuropathol Exp Neurol 57:369–384

79. Lawrence A, Hodges J, Rosser A, Kershaw A, French-Constant C, Rubinsztein D, Robbins T, Sahakian B (1998) Evidence for specific cognitive deficits in preclinical Huntington's disease. Brain Pathol 121:1329–1341

80. Kirkwood S, Siemers E, Hodes M, Conneally P, Christian J, Foroud T (2000) Subtle changes among presymptomatic carriers of the Huntington's disease gene. J Neurol Neurosurg Psychiatry 69:773–779

81. Hodgson J, Agopyan N, Gutekunst C, Leavitt B, LePiane F, Singaraja R, Smith D, Bissada N, McCutcheon K, Nasir J et al (1999) A YAC mouse model for Huntington's disease with full-length mutant huntingtin, cytoplasmic toxicity, and selective striatal neurodegeneration. Neuron 23:181–192

82. Usdin MT, Shelbourne PF, Myers RM, Madison DV (1999) Impaired synaptic plasticity in mice carrying the Huntington's disease mutation. Hum Mol Genet 8:839–846

83. Murphy KP, Carter RJ, Lione LA, Mangiarini L, Mahal A, Bates GP, Dunnett SB, Morton AJ (2000) Abnormal synaptic plasticity and impaired spatial cognition in mice transgenic for exon 1 of the human Huntington's disease mutation. J Neurosci 20:5115–5123

84. Zuccato C, Cattaneo E (2007) Role of brain-derived neurotrophic factor in Huntington's disease. Prog Neurobiol 81:294–330

85. Lauterborn J, Troung G, Baudry M, Bi X, Lynch G, Gall C (2003) Chronic elevation of brain-derived neurotrophic factor by ampakines. J Pharmacol Exp Ther 307:297–305

86. Lauterborn JC, Pineda E, Chen LY, Ramirez EA, Lynch G, Gall CM (2009) Ampakines cause sustained increases in brain-derived neurotrophic factor signaling at excitatory synapses without changes in AMPA receptor subunit expression. Neuroscience 159:283–295

87. Fedulov V, Rex CS, Simmons DA, Palmer L, Gall CM, Lynch G (2007) Evidence that long-term potentiation occurs within individual hippocampal synapses during learning. J Neurosci 27:8031–8039

88. Park D, Lautenschlager G, Hedden T, Davidson N, Smith A, Smith P (2002) Models of visuospatial and verbal memory across the adult life span. Psychol Aging 17:299–320

89. Deupree DL, Bradley J, Turner DA (1993) Age-related alterations in potentiation in the CA1 region in F344 rats. Neurobiol Aging 14:249–258

90. Herndon JG, Moss MB, Rosene DL, Killiany RJ (1997) Patterns of cognitive decline in aged rhesus monkeys. Behav Brain Res 87:25–34

91. Rex C, Kramar E, Colgin L, Lin B, Gall C, Lynch G (2005) Long-term potentiation is impaired in middle-aged rats: regional specificity and reversal by adenosine receptor antagonists. J Neurosci 25:5956–5966

92. Rex CS, Lauterborn JC, Lin CY, Kramar EA, Rogers GA, Gall CM, Lynch G (2006) Restoration of long-term potentiation in middle-aged hippocampus after induction of brain-derived neurotrophic factor. J Neurophysiol 96:677–685

93. Cunha R, Almeida T, Ribeiro J (2001) Parallel modification of adenosine extracellular metabolism and modulatory action in the hippocampus of aged rats. J Neurochem 76:372–382

94. Phillips S, Sherwin B (1993) Effects of estrogen on memory function in surgically menopausal women. Psychoneuroendocrinology 17:485–495

95. Devi G, Hahn K, Massimi S, Zhivotovskaya E (2005) Prevalence of memory loss complaints and other symptoms associated with the menopause transition: a community survey. Gend Med 2:255–264

96. Hachul H, Bittencourt L, Soares JJ, Tufik S, Baracat E (2009) Sleep in post-menopausal women: differences between early and late post-menopause. Eur J Obstet Gynecol Reprod Biol 145:81–84

97. Weber M, Mapstone M (2009) Memory complaints and memory performance in the menopausal transition. Menopause 16:694–700

98. Hojo Y, Murakami G, Mukai H, Higo S, Hatanaka Y, Ogiue-Ikeda M, Ishii H, Kimoto T, Kawato S (2008) Estrogen synthesis in the brain – role in synaptic plasticity and memory. Mol Cell Endocrinol 290:31–43

99. Kramar EA, Chen LY, Lauterborn JC, Simmons DA, Gall CM, Lynch G BDNF and BDNF up-regulation rescue synaptic plasticity in ovariectomized rats (Submitted)

100. Comery T, Harris J, Willems P, Oostra B, Irwin S, Weiler I, Greenough W (1997) Abnormal dendritic spines in fragile X knockout mice: maturation and pruning deficits. Proc Natl Acad Sci USA 94:5401–5404

101. Irwin S, Idupulapati M, Gilbert M, Harris J, Chakravarti A, Rogers E, Crisostomo R, Larsen B, Mehta A, Alacantara C et al (2002) Dendritic spine and dendritic field characteristics on layer V pyramidal neurons in the visual cortex of fragile-X knockout mice. Am J Med Genet 111:140–146

102. Purpura DP (1974) Dendritic spine "dysgenesis" and mental retardation. Science 186:1126–1128

103. Dindot S, Antalffy B, Bhattacharjee M, Beaudet A (2008) The Angelman syndrome ubiquitin ligase localizes to the synapse and nucleus, and maternal deficiency results in abnormal dendritic spine morphology. Hum Mol Genet 17:111–118

104. Penagarikano O, Mulle J, Warren S (2007) The pathophysiology of fragile x syndrome. Annu Rev Genomics Hum Genet 8:109–129

105. Dictenberg J, Swanger S, Antar L, Singer R, Bassell G (2008) A direct role for FMRP in activity-dependent dendritic mRNA transport links filopodial-spine morphogenesis to fragile X syndrome. Dev Cell 14:926–939

106. Davidovic L, Jaglin XH, Lepagnol-Bestel AM, Tremblay S, Simonneau M, Bardoni B, Khandjian EW (2007) The fragile X mental retardation protein is a molecular adaptor between the neurospecific KIF3C kinesin and dendritic RNA granules. Hum Mol Genet 16:3047–3058

107. Ling S, Fahrner P, Greenough W, Gelfand V (2004) Transport of Drosophila fragile X mental retardation protein-containing ribonucleoprotein granules by kinesin-1 and cytoplasmic dynein. Proc Natl Acad Sci USA 101:17428–17433

108. Ohashi S, Koike K, Omori A, Ichinose S, Ohara S, Kobayashi S, Sato TA, Anzai K (2002) Identification of mRNA/protein (mRNP) complexes containing Puralpha, mStaufen, fragile X protein, and myosin Va and their association with rough endoplasmic reticulum equipped with a kinesin motor. J Biol Chem 277:37804–37810

109. Lauterborn JC, Rex CS, Kramar E, Chen LY, Pandyarajan V, Lynch G, Gall CM (2007) Brain-derived neurotrophic factor rescues synaptic plasticity in a mouse model of fragile X syndrome. J Neurosci 27:10685–10694

110. Meredith RM, Holmgren CD, Weidum M, Burnashev N, Mansvelder HD (2007) Increased threshold for spike-timing-dependent plasticity is caused by unreliable calcium signaling in mice lacking fragile X gene FMR1. Neuron 54:627–638

111. Larson J, Jessen R, Kim D, Fine A, du Hoffmann J (2005) Age-dependent and selective impairment of long-term potentiation in the anterior piriform cortex of mice lacking the fragile X mental retardation protein. J Neurosci 25:9460–9469

112. Li J, Pelletier MR, Perez Velazquez JL, Carlen PL (2002) Reduced cortical synaptic plasticity and GluR1 expression associated with fragile X mental retardation protein deficiency. Mol Cell Neurosci 19:138–151

113. Hayashi ML, Rao BS, Seo JS, Choi HS, Dolan BM, Choi SY, Chattarji S, Tonegawa S (2007) Inhibition of p21-activated kinase rescues symptoms of fragile X syndrome in mice. Proc Natl Acad Sci USA 104:11489–11494

114. Zhao M, Toyoda H, Ko S, Ding H, Wu L, Zhuo M (2005) Deficits in trace fear memory and long-term potentiation in a mouse model for fragile X syndrome. J Neurosci 25:7385–7392

115. Bozdagi O, Rich E, Tronel S, Sadahiro M, Patterson K, Shapiro ML, Alberini CM, Huntley GW, Salton SR (2008) The neurotrophin-inducible gene Vgf regulates hippocampal function and behavior through a brain-derived neurotrophic factor-dependent mechanism. J Neurosci 28:9857–9869

116. Marty S, da Penha BM, Berninger B (1997) Neurotrophins and activity-dependent plasticity of cortical interneurons. Trends Neurosci 20:202

117. Kramar EA, Lin B, Rex CS, Gall CM, Lynch G (2006) Integrin-driven actin polymerization consolidates long-term potentiation. Proc Natl Acad Sci USA 103:5579–5584

118. Wang XB, Bozdagi O, Nikitczuk JS, Zhai ZW, Zhou Q, Huntley GW (2008) Extracellular proteolysis by matrix metalloproteinase-9 drives dendritic spine enlargement and long-term potentiation coordinately. Proc Natl Acad Sci USA 105:19520–19525

119. Kramar EA, Chen LY, Brandon NJ, Rex CS, Liu F, Gall CM, Lynch G (2009) Cytoskeletal changes underlie estrogen's acute effects on synaptic transmission and plasticity. J Neurosci 29:12982–12993

120. Chen LY, Rex CS, Babayan AH, Kramar EK, Lynch G, Gall CM, Lauterborn JC (2010) Physiological activation of synaptic Rac > PAK signaling is defective in a mouse model of fragile-X syndrome. J Neurosci (in press)

Activation of Group II Metabotropic Glutamate Receptors (mGluR2 and mGluR3) as a Novel Approach for Treatment of Schizophrenia

Douglas J. Sheffler and P. Jeffrey Conn

Abstract In the last decade, many advances have suggested that activators of mGluR2 and/or mGluR3 may provide a novel approach for the treatment of schizophrenia. Preclinical and clinical studies with traditional orthosteric agonists nonselective for mGluR2 and mGluR3 have provided key studies demonstrating the viability of this approach. Recent advances in allosteric ligand development have led to the discovery of positive allosteric modulators (PAMs) selective for mGluR2, which have demonstrated efficacy in animal models predictive of antipsychotic activity. These mGluR2 PAMs, in conjunction with studies in mGluR2, mGluR3, and mGluR2/3 knockout mice, have suggested that mGluR2 may be the predominant receptor responsible for the antipsychotic efficacy of nonselective mGluR2/3 agonists. Recently, a heterocomplex between mGluR2 and the 5-HT$_{2A}$ serotonin receptor has been reported with functional consequences for both receptors, providing a novel target for therapeutic development. Together, these many advances have provided a strong foundation for the continued development of mGluR2/3 activators as a novel approach for the treatment of schizophrenia.

1 Introduction

Schizophrenia is a severe psychiatric disorder affecting approximately 1% of the population worldwide that shows a similar prevalence in multiple cultures, geographies, and with no preference for gender [1]. This debilitating lifelong disorder has an onset in early adulthood, requiring daily maintenance therapy, with an estimated societal cost of approximately $65 billion annually (2002 data) [2]. Schizophrenic patients exhibit three core clusters of symptoms categorized as

P.J. Conn (✉)

Dept. of Pharmacology, Vanderbilt University Medical Center, Light Hall (MRB-IV) Room 1215D, 2215 B Garland Avenue, Nashville, TN 37232, USA

e-mail: jeff.conn@vanderbilt.edu

P. Skolnick (ed.), *Glutamate-based Therapies for Psychiatric Disorders*,
Milestones in Drug Therapy, DOI 10.1007/978-3-0346-0241-9_6,
© Springer Basel AG 2010

positive symptoms (delusions, paranoia, hallucinations, and catatonic behavior), negative symptoms (social withdrawal, anhedonia, apathy, and paucity of speech), and cognitive impairments (deficits in attention, working memory, and executive functions) [1]. The cognitive deficits are one of the major disabilities associated with schizophrenia and are considered a reliable predictor of long-term disability and treatment outcome [3–6]. Furthermore, patients exhibiting significant negative symptoms have a particularly poor functional capacity and quality of life [7]. Although the etiology of schizophrenia remains unknown, it is thought to be produced by a complex interaction between biological, environmental, and genetic factors. The absence of clear neuropathological changes responsible for this disorder has provided a challenge for the development of novel therapeutic approaches.

The first generation of antipsychotic drugs that were developed is collectively referred to as the typical antipsychotics. These drugs, typified by chlorpromazine and haloperidol, are efficacious for the treatment of positive symptoms. However, the typical antipsychotics have little efficacy for treatment of negative symptoms or cognitive deficits and may worsen cognitive deficits in some patients [8]. Typical antipsychotics also have a high burden of extrapyramidal side effects (EPS) including tardive dyskinesia and elevate serum prolactin levels, which severely limits their therapeutic use. Alternatively, the second generation atypical antipsychotics, typified by clozapine, display a relative lack of EPS and serum prolactin elevation compared with the typical antipsychotics. As with the typical antipsychotics, the atypicals display their primary efficacy toward the positive symptom cluster. Therefore, all currently available antipsychotic drugs are effective in treatment of the positive symptoms, but provide no or only modest effects on the negative symptoms and cognitive impairments. Furthermore, some patients are unresponsive to current antipsychotic treatments and several of these agents are associated with adverse effects, including disturbances in motor function, metabolic syndrome, weight gain, and sexual dysfunction [4, 5, 9, 10]. Clozapine, the most efficacious atypical antipsychotic available, also has additional liabilities including an increased risk of seizures and the development of agranulocytosis in 1–2% of patients [11]. Due to these limitations, it is imperative to develop new therapeutic strategies for schizophrenia that provide efficacy across all of the core symptom clusters and have fewer adverse effect liabilities.

Until recently, the dominant hypothesis of schizophrenia has been that excessive dopaminergic neurotransmission in the forebrain, particularly in the mesolimbic and striatal brain regions, leads to the positive symptoms, whereas dopaminergic deficits in prefrontal brain regions lead to the negative symptoms of schizophrenia [6, 12]. This dopaminergic hypothesis is based on the observation that all currently approved antipsychotics, both typical and atypical, display affinity toward D_2 dopamine receptors [13] where they either act as antagonists or partial agonists. For the typical antipsychotics in particular, the clinical efficacy of these compounds is correlated with their affinity for striatal D_2 receptors [14]. The dopaminergic hypothesis is supported by the observed alterations on striatal dopamine release in schizophrenic patients and is consistent with the psychotomimetic properties of indirect dopamine agonists such as amphetamine or cocaine [15–18]. However, at least 25% of patients do not respond to dopamine antagonists [19]. In addition, the therapeutic onset for D_2 antagonists is slow, implying that therapeutic efficacy results from an alteration of central nervous system (CNS) circuitry following chronic D_2 antagonist treatment and is not the result of acute receptor binding [20].

Atypical antipsychotics, in addition to their affinity at D_2 receptors, have high affinities at other biogenic amine receptors including the $5\text{-}HT_{2A}$ serotonin receptor, where they display potent antagonist activity. This has led to the hypothesis that atypical antipsychotics as a class can be distinguished by their $5\text{-}HT_{2A}/D_2$ affinity ratios, whereby atypical antipsychotic agents have lower affinities for D_2 receptors and higher affinities for $5\text{-}HT_{2A}$ receptors than do typical antipsychotics [21]. The affinity of the atypical antipsychotics for $5\text{-}HT_{2A}$ receptors led to the $5\text{-}HT_{2A}$ hyperfunction hypothesis of schizophrenia, which is supported by the psychoto-mimetic effects of $5\text{-}HT_{2A}$ agonists such as lysergic acid diethylamide (LSD). However, selective $5\text{-}HT_{2A}$ receptor antagonists such as (R)-(2,3-dimethoxyphenyl)-[1-[2-(4-fluorophenyl)ethyl]piperidin-4-yl]methanol (M100907) failed in a Phase III clinical trial for the treatment of acute schizophrenia, suggesting that selective $5\text{-}HT_{2A}$ antagonist monotherapy may not prove a sufficient therapeutic for schizophrenia [22]. As neither the dopamine hyperfunction hypothesis nor the serotonin hyperfunction hypothesis can fully account for the pathophysiology of schizophrenia and there are clear limitations of the currently available antipsychotics, there has been a tremendous research effort to identify additional neurochemical or neuro-physiological alterations that may contribute to schizophrenia's pathophysiology.

2 The Glutamatergic Hypothesis of Schizophrenia

Glutamate is the major excitatory neurotransmitter in the CNS and is responsible for the generation of fast excitatory synaptic responses at the vast majority of CNS synapses [23]. The fast excitatory synaptic responses at glutamatergic synapses are mediated by activation of the ionotropic glutamate receptors (iGluRs), which are comprised of the α-amino-3-hydroxy-5-methyl-4-isoazolepropionic acid (AMPA), N-methyl-D-aspartate (NMDA), and kainate receptor subtypes. These iGluRs are multimeric ligand-gated ion channels that open in response to glutamate, inducing excitatory postsynaptic currents [23]. Two dissociative anesthetics, phencyclidine (PCP) and ketamine, have been known for nearly half a century to induce a psychotic state [24]. The discovery that the mechanism of action of these drugs is through a noncompetitive blockade of NMDA receptors first implicated the gluta-matergic system in the generation of a psychotic state.

Early clinical findings have also demonstrated that NMDA receptor antagonists cause psychotic symptoms in healthy humans that are difficult to distinguish from schizophrenia in the clinic and exacerbate symptoms in schizophrenic patients [25, 26]. These studies provided the basis of a potential link between NMDA recep-tor signaling and schizophrenia. The finding that the psychotic state induced by NMDA receptor antagonists precisely mimics the positive, negative, and cognitive symptom clusters of schizophrenia suggests that the dopaminergic dysfunction in schizophrenia is secondary to an underlying glutamatergic dysfunction. Several studies in schizophrenic patients further support a role for altered glutamatergic neurotransmission, including a decrease in glutamate concentration in cerebrospinal fluid that inversely correlates with the severity of positive symptoms [27], a decrease in NMDA-induced glutamate release in synaptosomes prepared from schizophrenic brains [28], and alterations in NMDA receptor density in a variety of regions in the schizophrenic brain [29–31]. Together these studies form the basis of the NMDA receptor hypofunction hypothesis of schizophrenia and suggest that agents that increase NMDA receptor function may be efficacious for the treatment of all of schizophrenia's symptom clusters [32–35]. One highly attractive feature of the glutamatergic hypothesis of schizophrenia is that it shares many consistencies with the dopaminergic hyperfunction hypothesis. For instance, NMDA antagonists increase cortical dopamine efflux, suggesting NMDA hypofunction may be a causative factor in inducing a hyperdopaminergic state. This effect may in part explain the efficacy of typical and atypical antipsychotics in reversing the psychotomimetic effects of NMDA receptor antagonists [35], although far more work is necessary to fully understand these effects. Given the implication of a hypofunction of NMDA receptors in schizophrenia, on the surface, NMDA receptors themselves would make an attractive target for drug development. However, direct nonselective NMDA receptor agonists would be expected to produce numerous unwanted side effects including cognitive impairment, seizure activity, and excitotoxicity. Due to these limitations, there has been a quest to identify more indirect methods to augment NMDA receptor function.

Several animal models of NMDA receptor hypofunction-induced behaviors have been critical for the preclinical evaluation of novel therapeutic targets. Administration of NMDA receptor antagonists to rodents elicits numerous behavioral effects similar to schizophrenia including deficits in prepulse inhibition (PPI) (a measure of sensorimotor gating), increases in locomotor activity, and cognitive deficits [36]. These effects are attenuated by both typical and atypical antipsychotic treatment [34]. Therefore, these rodent behavioral models provide valuable systems for prediction of antipsychotic drug action. Consequently, the majority of drug discovery efforts are aimed at the ability of novel compounds to reverse the behavioral effects of NMDA antagonists in these models [37]. In further support of these behavioral models, the generation of a mutant mouse expressing only 5% of normal levels of the NMDA receptor NR1 subunit has shown that decreasing functional NMDA receptors induces a similar state to that seen in other animal models of schizophrenia. Importantly, the behavioral abnormalities in these rodents are attenuated by typical or atypical antipsychotic treatment [20]. The advent of these rodent models has provided necessary in vivo tools for the evaluation of novel antipsychotics.

More recently, studies have demonstrated that a reduction in NMDA receptor function induces complex changes in cortical and subcortical circuits of both glutamatergic and GABAergic synapses. For example, although there is a general decrease in glutamate found in cerebrospinal fluid, blockade of NMDA receptors with PCP causes increased glutamate efflux in the prefrontal cortex and the nucleus accumbens [38]. Interestingly, these circuitry alterations may share common features with changes found in schizophrenic patients [39]. This effect of NMDA receptor blockade is hypothesized to be mediated via a selective decrease in the activity of GABAergic interneurons compared with glutamatergic neurons, leading to the disinhibition of downstream glutamatergic neurons and an enhancement of glutamate release [38, 40–42]. Consistent with this hypothesis, a decrease in GABAergic interneurons in the anterior cingulate cortex, prefrontal cortex, and hippocampus has been seen in the schizophrenic brain [43, 44].

In a model of this neuronal circuitry, GABAergic projection neurons in subcortical regions such as the nucleus accumbens normally provide inhibitory control onto glutamatergic thalamocortical neurons projecting to the prefrontal cortex. Disruption of this inhibitory control, which could occur through a hypofunction of NMDA receptors on GABAergic projection neurons (Fig. 1a), could in turn result in disinhibition of thalamocortical inputs to pyramidal neurons in the

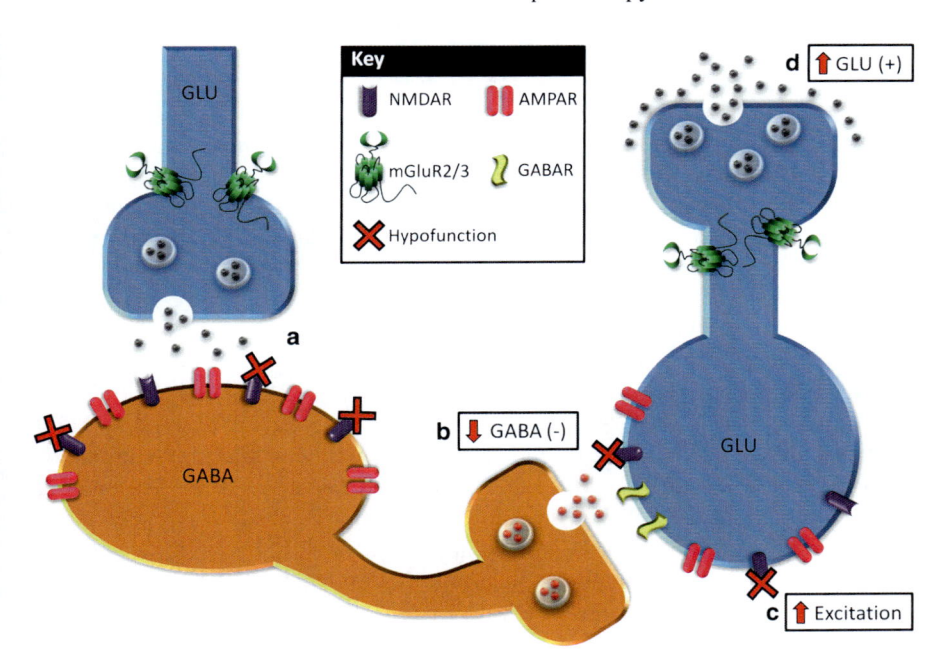

Fig. 1 NMDA receptor hypofunction model for schizophrenia. (**a**) Normally, NMDA receptor activation on GABAergic projection neurons (*orange*) in subcortical regions such as the nucleus accumbens provides inhibitory control onto glutamatergic thalamocortical neurons projecting to the prefrontal cortex. (**b**) NMDA receptor hypofunction on these GABAergic neurons could result in a disruption of inhibitory control. (**c**) Decreased GABAergic projection neuron excitation could result in disinhibition of thalamocortical inputs to pyramidal neurons in the prefrontal cortex, leading to increased excitation of thalamic glutamatergic neurons mediated via AMPA receptors. (**d**) This would result in subsequent increased glutamate efflux at thalamocortical synapses in the prefrontal cortex. mGluR2/3 receptors are expressed in key regions for modulation of this altered circuitry

prefrontal cortex (Fig. 1b). This disinhibition would lead to increased excitation of thalamic glutamatergic neurons mediated via non-NMDA glutamate receptors such as AMPA receptors (Fig. 1c) and a subsequent increased glutamate efflux at thalamocortical synapses in the prefrontal cortex (Fig. 1d). This model has led to the hypothesis that reduction of excitatory glutamatergic neurotransmission at key synapses, such as thalamocortical synapses in the prefrontal cortex, may provide a novel therapeutic approach for antipsychotic drug development. One possible mechanism by which this may be accomplished is through the activation of the metabotropic glutamate receptors subtypes 2 and 3 (mGluR2 and mGluR3), which are presynaptically localized in these synapses where they are poised to inhibit glutamatergic neurotransmission (Fig. 1) [3, 32, 34, 35, 45–48].

3   The Group II mGluRs as Therapeutics Targets for Schizophrenia

In addition to the activation of iGluRs outlined above, glutamate activates metabotropic glutamate receptors (mGluRs), which are G protein-coupled receptors (GPCRs) [49]. These family C GPCRs have a large extracellular domain which comprises the orthosteric binding site for their endogenous ligand glutamate, seven transmembrane domains connected by three intracellular and three extracellular loops, and an intracellular C-terminal domain (Fig. 2). The mGluRs provide a mechanism by which glutamate can modulate activity at the same synapses at which it elicits fast synaptic responses via the iGluRs [50]. Eight receptor subtypes of mGluRs have been cloned, which are delineated based upon G protein-coupling specificity, pharmacology, and sequence homology [51]. Group I mGluRs (mGluR1 and mGluR5) are traditionally coupled to $G_{\alpha q}$, leading to the stimulation of phospholipase Cβ, phosphoinositide hydrolysis, and intracellular calcium mobilization. Group II mGluRs (mGluR2 and mGluR3) and Group III mGluRs (mGluR4, mGluR6, mGluR7, and mGluR8) are coupled to $G_{\alpha i/o}$-associated signal transduction cascades and can also regulate ion channel function via $G_{\beta \gamma}$ subunit activity [51]. Given the ubiquitous distribution of glutamatergic synapses and the differential expression mGluRs at these synapses, mGluRs participate in a wide variety of functions of the CNS. Thus, the opportunity exists for developing therapeutic agents that selectively interact with mGluRs involved in only one or a limited number of CNS functions [50, 52, 53].

A number of preclinical and clinical studies provide evidence that agonists of the Group II mGluRs have potential as a novel therapeutic approach for schizophrenia. Expression studies, including mRNA, protein, and autoradiography studies, have shown that mGluR2 and mGluR3 are highly expressed in many regions of the rat brain relevant to schizophrenia, including the hippocampus, striatum, amygdala, nucleus accumbens, cerebral cortex, and the cerebellar cortex [34, 54]. These receptors are predominantly presynaptic, where they inhibit glutamate release

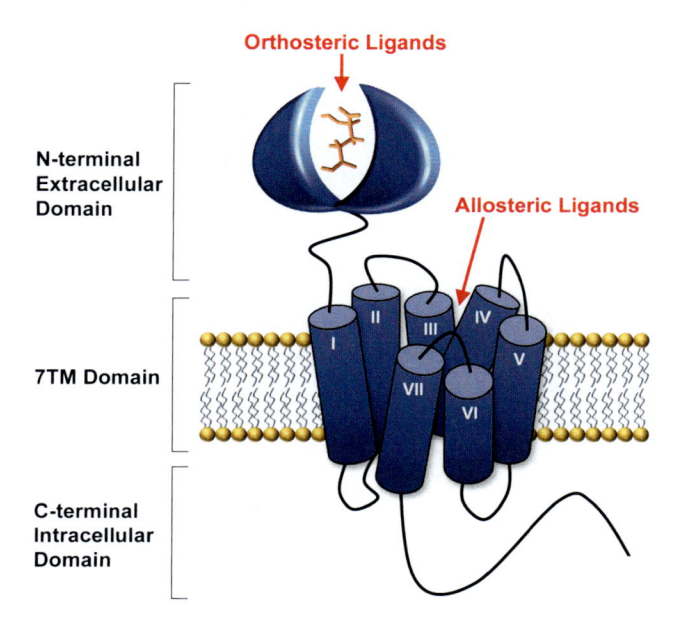

Fig. 2 Schematic illustration of the mGluR structure. mGluRs possess a large N-terminal extracellular domain that contains the orthosteric binding site of the endogenous ligand glutamate, seven putative transmembrane (7TM) domains connected via three intracellular and three extracellular loops, and a cytoplasmic C-terminal tail. Allosteric ligands bind to sites other than the orthosteric glutamate-binding site, such as within the 7TM domain

upon activation, reducing postsynaptic excitation. In addition to their role in glutamatergic neurotransmission, the Group II mGluRs have also been shown to function as heteroreceptors acting to inhibit GABA release [55]. In opposition to their largely overlapping neuronal expression, mGluR3, but not mGluR2, has a high level of expression in glial cells throughout the brain [56, 57]. Importantly, activation of presynaptic group II mGluRs reduces transmission in several areas relevant to schizophrenia including at the medial perforant path-dentate gyrus synapse [58], the mossy fiber synapse [59], and at synapses onto certain interneuron populations [60]. Group II mGluR agonists have also been shown to inhibit excitatory post synaptic potentials (EPSPs) in cortical areas, cortico-striatal, dopaminergic midbrain, and central and basolateral amygdala neurons [34]. Furthermore, Group II mGluRs also play a modulatory role on thalamocortical glutamatergic neurotransmission. As outlined above and shown in Fig. 1, psychotomimetics increase glutamatergic synapse activity in the medial prefrontal cortex, which is postulated to play a role in the pathophysiology of schizophrenia [34, 61, 62]. Interestingly, the effects of psychotomimetics in this region are blocked by Group II agonists [63].

A number of nonselective orthosteric mGluR2/3 agonists have been developed which are conformationally constrained amino acid analogs. These include (1S,2S, 5R,6S)-2-aminobicyclo-[3.1.0]-hexane-2,6-dicarboxylate (LY354740), (1S,2R, 5R,6R)-2-amino-4-oxabicyclo-[3.1.0]-hexane-2,6-dicarboxylate (LY379268), and

(-)-(1R,4S,5S,6S)-4-amino-2-sulfonylbicyclo-[3.1.0]-hexane-4,6-dicarboxylate (LY404039), whose structures are shown in Fig. 3. In addition, a reasonably selective orthosteric antagonist of mGluR2/3 has been developed, 2-[(1S,2S)-2-carboxycyclopropyl]-3-(9H-xanthen-9-yl)-D-alanine (LY341495), which has provided a valuable research tool for investigating the effects of the Group II agonists. These orthosteric agonists provided the first initial data that suggest mGluR2/3 are valuable targets for antipsychotic drug development as they have been demonstrated to have antipsychotic activity in numerous animal models used to predict antipsychotic efficacy [4, 54]. For example, the orthosteric agonists LY354740 and LY379268 attenuate PCP-induced hyperlocomotor activity, stereo-typed behaviors, and working memory deficits in rats [62, 64] and the effects of these Group II agonists are blocked by the Group II antagonist LY341495 [64]. Furthermore, these compounds also inhibit PCP [62] or ketamine-induced [61] glutamate release in prefrontal cortex and nucleus accumbens. These preclinical studies validated the use of mGluR2/3 activators as potential antipsychotics and paved the way for evaluation of this approach in clinical models.

Clinical studies have also demonstrated antipsychotic efficacy of orthosteric mGluR2/3 agonists, a key finding for this therapeutic approach. An early clinical study with LY354740 demonstrated an improvement in ketamine-induced cognitive deficits in healthy subjects [65]. More recently, as a proof-of-concept for this potential therapeutic class, LY2140023, the prodrug of LY404039, became the first clinically tested nondopamine antagonist schizophrenia drug. In this clinical study, schizophrenic patients showed improvements in positive and negative symptoms over the course of the trial. Furthermore, the effects found were comparable to atypical antipsychotic olanzapine but without the side effect profile of the typical

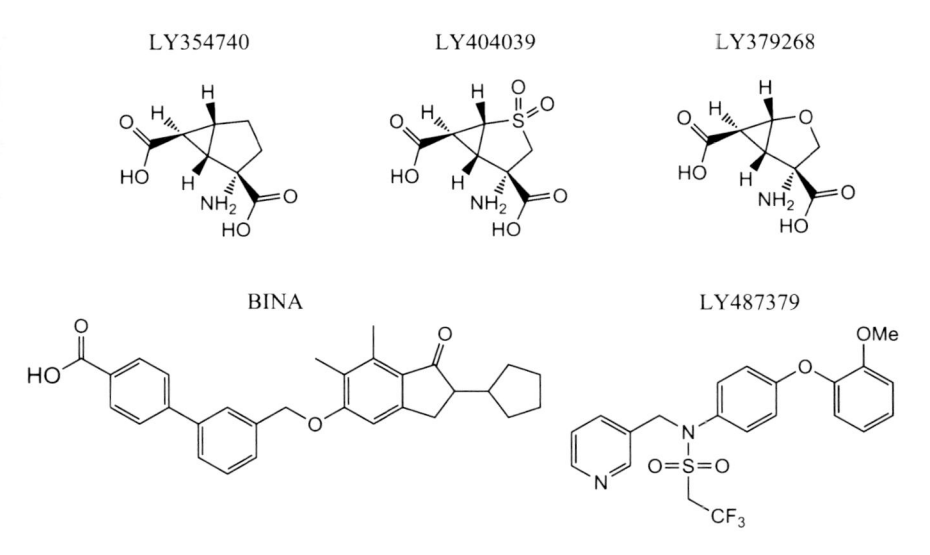

Fig. 3 Representative chemical structures of Group II mGluR orthosteric agonists (LY354740, LY379268, and LY404039) and mGluR2 PAMs (BINA and LY487379)

or atypical antipsychotics. Encouragingly, LY2140023 did not elevate serum prolactin, cause EPS, or weight gain in patients [66]. However, despite these advances, several drawbacks call into question the long-term clinical utility of Group II mGluR orthosteric agonists. All of the orthosteric agonists developed thus far are centered on one major chemical scaffold, namely, they are all conformationally constrained amino acid analogs (see Fig. 3). Given this lack of diversity in the chemical space, improvements over the currently available orthosteric agonists may not be possible. Unfortunately, robust tolerance of orthosteric agonists may prove to be an additional liability as repeated dosing of LY379268 led to the development of tolerance for its ability to block PCP or amphetamine-induced hyperlocomotor activity in rats [67]. Furthermore, as these Group II orthosteric agonists are nonselective for mGluR2 or mGluR3, studies with these compounds do not provide insight into whether mGluR2 or mGluR3 is more important for their clinical efficacy. Although orthosteric agonists selective for either mGluR2 or mGluR3 have not been available, mGluR2, mGluR3, and mGluR2/3 knockout mice have been used as valuable model systems to determine the individual roles of the Group II mGluRs in the antipsychotic effects of Group II agonists. For example, the Group II orthosteric agonists LY404039 and LY379268 reverse PCP or amphetamine-induced hyperlocomotion in mGluR3 knockout mice but not in mGluR2 knockout mice [68, 69]. These data first suggested that mGluR2 may be the predominant receptor responsible for the antipsychotic efficacy of nonselective mGluR2/3 agonists.

4 mGluR2 Positive Allosteric Modulators as Therapeutics for Schizophrenia

The studies outlined above provide support for the hypothesis that mGluR2 and/or mGluR3 activators may provide a novel approach for the treatment of schizophrenia. However, the initial efforts to identify mGluR subtype-selective ligands have focused on the development of competitive orthosteric site ligands, namely, ligands that bind to the endogenous glutamate-binding site present in the large N-terminal extracellular domain of the mGluRs (Fig. 2). Unfortunately, it has proven difficult to develop highly selective ligands that act at the glutamate-binding site, likely due to the high conservation of the orthosteric binding site across mGluR subtypes [52]. Another approach that has been highly successful in the clinical setting is the use of selective positive allosteric modulators (PAMs) of specific receptor subtypes. These small molecules do not activate the mGluRs directly but act at allosteric sites on the receptor to potentiate glutamate-induced receptor activation (Fig. 2). The advantage of this approach is that putative allosteric sites are less highly conserved, affording greater opportunity for the identification of subtype-selective ligands. Furthermore, mGluR PAMs only potentiate the activity of endogenously released glutamate, a feature that should maintain activity dependence of receptor activation. As the receptor will only be potentiated when endogenous glutamate is released, this may allow for less receptor desensitization to occur and may potentially minimize side effect profiles.

Several highly selective mGluR2 PAMs have been developed, the majority of which are related to two structural classes represented by N-(4-(2-methoxyphenoxy)phenyl)-N-(2,2,2-trifluoroethylsulfonyl)pyrid-3-ylmethylamine (LY487379, 4-MPPTS) [67, 70] and 3′-(((2-cyclopentyl-6,7-dimethyl-1-oxo-2,3-dihydro-1H-inden-5-yl)oxy)methyl)biphenyl-4-carboxylic acid (Biphenylindanone A, BINA) [71–73], whose structures are shown in Fig. 3. These compounds are highly selective for mGluR2 and do not potentiate responses to the activation of mGluR3 or any other mGluR subtype. In most systems, the mGluR2 PAMs show no agonist activity but induce leftward shifts in the concentration-response curve to glutamate. Consistent with their allosteric mechanism of action, mutational analysis has shown that three amino acids in the 7TM domain, residing in TMIV and TMV, are critical for the activity of mGluR2 PAMs [74, 75], a site far removed from the orthosteric binding site.

The mGluR2 PAMs have been demonstrated to have effects in potentiating Group II mGluR agonist responses at several glutamatergic synapses [70, 73, 74, 76, 77] including within the prefrontal cortex, where their effects are thought to be relevant for therapeutic benefit in schizophrenia [76]. Furthermore, multiple distinct mGluR2 PAMs have demonstrated efficacy in animal models predictive of antipsychotic activity and these effects are similar to antipsychotic activity of the mGluR2/3 orthosteric agonists [67, 70, 72, 73, 76, 78]. These studies further suggest that mGluR2 is the primary contributor to the efficacy of mGluR2/3 agonists in animal models of schizophrenia. Interestingly, the mGluR2 PAM LY487379 has been shown to reverse the amphetamine-induced disruption in PPI, whereas orthosteric agonists such as LY379268 have no effect [67], raising the possibility that mGluR2 selective activators may have efficacy toward the sensorimotor gating deficits associated with schizophrenia. Taken together, these studies raise the possibility that selective mGluR2 PAMs might provide a novel approach to the treatment of schizophrenia that may be devoid of the adverse effects of the currently available antipsychotic drugs.

5   mGluR2 and 5-HT$_{2A}$ May Provide a Novel Heterocomplex to Target for Potential Antipsychotic Activity

It has been known for some time that there is an association between the serotonergic and glutamatergic systems that may be of particular relevance for schizophrenia. Serotonin induces an increase in spontaneous excitatory postsynaptic currents (EPSCs) on the proximal dendrites of layer V pyramidal cells throughout the neocortex, leading to an increase in asynchronous glutamate release in the medial prefrontal cortex [79]. Further studies have demonstrated that 5-HT$_{2A}$ receptors mediate these increases in glutamate efflux via their actions on thalamocortical afferents [80].

Importantly, mGluR2/3 activators including LY354740, LY379268, and the mGluR2 PAM BINA have been shown to block 5-HT$_{2A}$-mediated excitatory postsynaptic currents in the rat prefrontal cortex [63, 76]. Group II mGluR activation also prevents the down-regulation of prefrontal cortical 5-HT$_{2A}$ receptors following repeated administration of 1-[2,5-dimethoxy-4-iodophenyl]-2-aminopropane (DOI), a 5-HT$_{2A}$ receptor hallucinogenic agonist [81]. It is hypothesized that these effects on 5-HT$_{2A}$ receptors may be one way by which the Group II agonists attenuate the effects of 5-HT$_{2A}$ hallucinogenic agonists such as LSD.

A valuable animal model for investigating the effects of hallucinogenic 5-HT$_{2A}$ receptor agonists is the stereotyped head twitch response in rodents. Hallucinogenic 5-HT$_{2A}$ receptor agonists induce this response, which is dependent on the 5-HT$_{2A}$ receptor since they are absent in 5-HT$_{2A}$ knockout mice [82]. Interestingly, both mGluR2/3 orthosteric agonists and the mGluR2 PAM BINA have been shown to attenuate this response, further supporting the targeting of mGluR2 as a novel strategy for treating glutamatergic dysfunction in schizophrenia [76]. In addition to these functional interaction studies between the glutamatergic and serotonergic systems, there is a high degree of overlap between the expression of 5-HT$_{2A}$ receptors and Group II mGluRs. Autoradiography studies have demonstrated an overlap between the Group II mGluRs and 5-HT$_{2A}$ receptors in the prefrontal cortex, particularly in the superficial and midlayers [63]. As outlined above, disruptions in the thalamocortical circuitry have been suggested to be altered in schizophrenia (Fig. 1). 5-HT$_{2A}$ antagonist modulation of this circuitry may be one mechanism by which the atypical antipsychotics provide therapeutic benefit in schizophrenia.

Recently, a heterodimeric complex between mGluR2 and the 5-HT$_{2A}$ serotonin receptor has been reported. This study demonstrated a direct molecular interaction between these two receptor subtypes, requiring TMIV and TMV of mGluR2 [83]. This interaction is proposed to occur between postsynaptic mGluR2 and 5-HT$_{2A}$ receptors, both of which are highly expressed by cortical pyramidal neurons. Importantly, the interaction between these receptors appears to have functional consequences for both receptors. In this study, mGluR2/3 agonists increased the affinity of hallucinogens at 5-HT$_{2A}$ receptors. Inversely, 5-HT$_{2A}$ agonists decreased the affinity of Group II agonists for mGluR2. Activation of mGluR2 in this complex also alters the signal transduction downstream of 5-HT$_{2A}$ receptors by inhibiting G$_{i/o}$ signal transduction pathways leading to the induction of the immediate early gene egr-2, a specific marker for 5-HT$_{2A}$-mediated hallucinogen signaling. Furthermore, this study found that 5-HT$_{2A}$ receptors are up-regulated and mGluR2 is down-regulated in the frontal cortex of patients with schizophrenia [83]. More recent studies have also demonstrated that activation of mGluR2/3 results in the in vivo inhibition of phosphoinositide hydrolysis mediated via 5-HT$_{2A}$ receptors. This study also demonstrated a negative interaction between 5-HT$_{2A}$ receptors and mGluR2/3 on the mitogen activated protein kinase (MAPK) pathway. Normally, activation of either 5-HT$_{2A}$ receptors or mGluR2/3 receptors results in an increase in phosphorylation of the MAPK cascade members extracellular signal-regulated kinases 1 and 2 (ERK1/2). However, when both 5-HT$_{2A}$ receptors and mGluR2/3 are coactivated, phosphorylation of ERK1/2 is highly attenuated, showing a reciprocal antagonism for this complex on the MAPK pathway [84]. Further elucidation of the relevance of the complex to the pathophysiology of schizophrenia will continue to integrate the functional interactions between the glutamatergic and serotonergic systems.

6 Concluding Remarks

There have been a number of advances in the last decade suggesting that activators of mGluR2 and/or mGluR3 may provide a novel approach for the treatment of schizophrenia. Preclinical and clinical studies with traditional orthosteric agonists nonselective for mGluR2 and mGluR3 have provided key data demonstrating the viability of this approach. The advent of allosteric ligand development for the Group II mGluRs has resulted in the development of several selective mGluR2 PAMs, which have been shown to have efficacy in a number of animal models predictive of antipsychotic drug action. These studies, combined with studies in mGluR2 and mGluR3 knockout mice, suggest that mGluR2 may be the primary mediator of the antipsychotic efficacy of mGluR2/3 nonselective agonists, though far more additional work is necessary to fully understand the individual roles of mGluR2 and mGluR3. Most importantly, selective reagents for mGluR3 are needed to begin to fully dissect the role of mGluR3 as a potential therapeutic target for schizophrenia. The recent identification of mGluR2/5-HT$_{2A}$ heterocomplexes and the studies aimed at understanding their functional relevance may provide yet another novel target for antipsychotic drug development. Together, these many advances have provided a strong foundation for the continued development of mGluR2/3 activators as a novel approach for the treatment of schizophrenia.

References

1. Lang UE et al (2007) Molecular mechanisms of schizophrenia. Cell Physiol Biochem 20 (6):687–702
2. Wu EQ et al (2005) The economic burden of schizophrenia in the United States in 2002. J Clin Psychiatry 66(9):1122–1129
3. Conn PJ, Lindsley CW, Jones CK (2009) Activation of metabotropic glutamate receptors as a novel approach for the treatment of schizophrenia. Trends Pharmacol Sci 30(1):25–31
4. Conn PJ et al (2008) Schizophrenia: moving beyond monoamine antagonists. Mol Interv 8 (2):99–107
5. Meltzer HY (1999) Treatment of schizophrenia and spectrum disorders: pharmacotherapy, psychosocial treatments, and neurotransmitter interactions. Biol Psychiatry 46(10): 1321–1327
6. Tan HY, Callicott JH, Weinberger DR (2007) Dysfunctional and compensatory prefrontal cortical systems, genes and the pathogenesis of schizophrenia. Cereb Cortex 17(Suppl 1): 171–181

7. Gray JA, Roth BL (2007) The pipeline and future of drug development in schizophrenia. Mol Psychiatry 12(10):904–922
8. Purdon SE et al (2001) Neuropsychological change in patients with schizophrenia after treatment with quetiapine or haloperidol. J Psychiatry Neurosci 26(2):137–149
9. Nikam SS, Awasthi AK (2008) Evolution of schizophrenia drugs: a focus on dopaminergic systems. Curr Opin Investig Drugs 9(1):37–46
10. Lamberti JS et al (2006) Prevalence of the metabolic syndrome among patients receiving clozapine. Am J Psychiatry 163(7):1273–1276
11. Alvir JM et al (1993) Clozapine-induced agranulocytosis. Incidence and risk factors in the United States. N Engl J Med 329(3):162–167
12. Toda M, Abi-Dargham A (2007) Dopamine hypothesis of schizophrenia: making sense of it all. Curr Psychiatry Rep 9(4):329–336
13. Miyamoto S et al (2003) Recent advances in the neurobiology of schizophrenia. Mol Interv 3 (1):27–39
14. Creese I, Burt DR, Snyder SH (1976) Dopamine receptor binding predicts clinical and pharmacological potencies of antischizophrenic drugs. Science 192(4238):481–483
15. Breier A et al (1998) Effects of NMDA antagonism on striatal dopamine release in healthy subjects: application of a novel PET approach. Synapse 29(2):142–147
16. Breier A et al (1997) Schizophrenia is associated with elevated amphetamine-induced synaptic dopamine concentrations: evidence from a novel positron emission tomography method. Proc Natl Acad Sci USA 94(6):2569–2574
17. Laruelle M et al (1996) Single photon emission computerized tomography imaging of amphetamine-induced dopamine release in drug-free schizophrenic subjects. Proc Natl Acad Sci USA 93(17):9235–9240
18. Laruelle M et al (1997) Imaging D2 receptor occupancy by endogenous dopamine in humans. Neuropsychopharmacology 17(3):162–174
19. Hirsch S, Barnes TRE (1995) The clinical treatment of schizophrenia with antipsychotic medication. In: Hirsch and Weinberger (eds) Schizophrenia, Blackwell Science,Oxford, UK, pp 443–468
20. Mohn AR et al (1999) Mice with reduced NMDA receptor expression display behaviors related to schizophrenia. Cell 98(4):427–436
21. Meltzer HY, Matsubara S, Lee JC (1989) Classification of typical and atypical antipsychotic drugs on the basis of dopamine D-1, D-2 and serotonin2 pKi values. J Pharmacol Exp Ther 251(1):238–246
22. de Paulis T (2001) M-100907 (Aventis). Curr Opin Investig Drugs 2(1):123–132
23. Dingledine R et al (1999) The glutamate receptor ion channels. Pharmacol Rev 51(1):7–61
24. Javitt DC, Zukin SR (1991) Recent advances in the phencyclidine model of schizophrenia. Am J Psychiatry 148(10):1301–1308
25. Krystal JH et al (1994) Subanesthetic effects of the noncompetitive NMDA antagonist, ketamine, in humans. Psychotomimetic, perceptual, cognitive, and neuroendocrine responses. Arch Gen Psychiatry 51(3):199–214
26. Shulgin AT (1964) 3-Methoxy-4, 5-methylenedioxy amphetamine, a new psychotomimetic agent. Nature 201:1120–1121
27. Faustman WO et al (1999) Cerebrospinal fluid glutamate inversely correlates with positive symptom severity in unmedicated male schizophrenic/schizoaffective patients. Biol Psychiatry 45(1):68–75
28. Sherman AD et al (1991) Deficient NMDA-mediated glutamate release from synaptosomes of schizophrenics. Biol Psychiatry 30(12):1191–1198
29. Akbarian S et al (1996) Selective alterations in gene expression for NMDA receptor subunits in prefrontal cortex of schizophrenics. J Neurosci 16(1):19–30
30. Gao XM et al (2000) Ionotropic glutamate receptors and expression of N-methyl-D-aspartate receptor subunits in subregions of human hippocampus: effects of schizophrenia. Am J Psychiatry 157(7):1141–1149

31. Law AJ, Deakin JF (2001) Asymmetrical reductions of hippocampal NMDAR1 glutamate receptor mRNA in the psychoses. NeuroReport 12(13):2971–2974

32. Coyle JT (2006) Glutamate and schizophrenia: beyond the dopamine hypothesis. Cell Mol Neurobiol 26(4–6):365–384

33. Lindsley CW et al (2006) Progress towards validating the NMDA receptor hypofunction hypothesis of schizophrenia. Curr Top Med Chem 6(8):771–785

34. Chavez-Noriega LE et al (2005) Novel potential therapeutics for schizophrenia: focus on the modulation of metabotropic glutamate receptor function. Curr Neuropharmacol 3(1):9–34

35. Marino MJ, Conn PJ (2002) Direct and indirect modulation of the N-methyl D-aspartate receptor. Curr Drug Targets CNS Neurol Disord 1(1):1–16

36. Kristiansen LV et al (2007) NMDA receptors and schizophrenia. Curr Opin Pharmacol 7(1):48–55

37. Spooren W et al (2003) Insight into the function of Group I and Group II metabotropic glutamate (mGlu) receptors: behavioural characterization and implications for the treatment of CNS disorders. Behav Pharmacol 14(4):257–277

38. Svensson TH (2000) Dysfunctional brain dopamine systems induced by psychotomimetic NMDA-receptor antagonists and the effects of antipsychotic drugs. Brain Res Brain Res Rev 31(2–3):320–329

39. Lisman JE et al (2008) Circuit-based framework for understanding neurotransmitter and risk gene interactions in schizophrenia. Trends Neurosci 31(5):234–242

40. Carlsson A et al (2001) Interactions between monoamines, glutamate, and GABA in schizophrenia: new evidence. Annu Rev Pharmacol Toxicol 41:237–260

41. Greene R (2001) Circuit analysis of NMDAR hypofunction in the hippocampus, in vitro, and psychosis of schizophrenia. Hippocampus 11(5):569–577

42. Tsai G, Coyle JT (2002) Glutamatergic mechanisms in schizophrenia. Annu Rev Pharmacol Toxicol 42:165–179

43. Benes FM et al (1998) A reduction of nonpyramidal cells in sector CA2 of schizophrenics and manic depressives. Biol Psychiatry 44(2):88–97

44. Benes FM, Vincent SL, Todtenkopf M (2001) The density of pyramidal and nonpyramidal neurons in anterior cingulate cortex of schizophrenic and bipolar subjects. Biol Psychiatry 50(6):395–406

45. Aghajanian GK, Marek GJ (2000) Serotonin model of schizophrenia: emerging role of glutamate mechanisms. Brain Res Brain Res Rev 31(2–3):302–312

46. Carlsson A et al (1997) Neurotransmitter aberrations in schizophrenia: new perspectives and therapeutic implications. Life Sci 61(2):75–94

47. Moreno JL, Sealfon SC, Gonzalez-Maeso J (2009) Group II metabotropic glutamate receptors and schizophrenia. Cell Mol Life Sci 66(23):3777–3785

48. Fraley ME (2009) Positive allosteric modulators of the metabotropic glutamate receptor 2 for the treatment of schizophrenia. Expert Opin Ther Pat 19(9):1259–1275

49. Pin JP, Duvoisin R (1995) The metabotropic glutamate receptors: structure and functions. Neuropharmacology 34(1):1–26

50. Anwyl R (1999) Metabotropic glutamate receptors: electrophysiological properties and role in plasticity. Brain Res Brain Res Rev 29(1):83–120

51. Nakanishi S (1994) Metabotropic glutamate receptors: synaptic transmission, modulation, and plasticity. Neuron 13(5):1031–1037

52. Conn PJ, Pin JP (1997) Pharmacology and functions of metabotropic glutamate receptors. Annu Rev Pharmacol Toxicol 37:205–237

53. Coutinho V, Knopfel T (2002) Metabotropic glutamate receptors: electrical and chemical signaling properties. Neuroscientist 8(6):551–561

54. Schoepp DD, Marek GJ (2002) Preclinical pharmacology of mGlu2/3 receptor agonists: novel agents for schizophrenia? Curr Drug Targets CNS Neurol Disord 1(2):215–225

55. Forsythe ID, Barnes-Davies M (1997) Synaptic transmission: well-placed modulators. Curr Biol 7(6):R362–R365

56. Ohishi H et al (1993) Distribution of the mRNA for a metabotropic glutamate receptor (mGluR3) in the rat brain: an in situ hybridization study. J Comp Neurol 335(2):252–266
57. Fotuhi M et al (1994) Differential expression of metabotropic glutamate receptors in the hippocampus and entorhinal cortex of the rat. Brain Res Mol Brain Res 21(3–4):283–292
58. Macek TA et al (1996) Differential involvement of group II and group III mGluRs as autoreceptors at lateral and medial perforant path synapses. J Neurophysiol 76(6):3798–3806
59. Nicholls RE et al (2006) mGluR2 acts through inhibitory Galpha subunits to regulate transmission and long-term plasticity at hippocampal mossy fiber-CA3 synapses. Proc Natl Acad Sci USA 103(16):6380–6385
60. Doherty JJ et al (2004) Metabotropic glutamate receptors modulate feedback inhibition in a developmentally regulated manner in rat dentate gyrus. J Physiol 561(Pt 2):395–401
61. Lorrain DS et al (2003) Effects of ketamine and N-methyl-D-aspartate on glutamate and dopamine release in the rat prefrontal cortex: modulation by a group II selective metabotropic glutamate receptor agonist LY379268. Neuroscience 117(3):697–706
62. Moghaddam B, Adams BW (1998) Reversal of phencyclidine effects by a group II metabotropic glutamate receptor agonist in rats. Science 281(5381):1349–1352
63. Marek GJ et al (2000) Physiological antagonism between 5-hydroxytryptamine(2A) and group II metabotropic glutamate receptors in prefrontal cortex. J Pharmacol Exp Ther 292 (1):76–87
64. Cartmell J, Monn JA, Schoepp DD (1999) The metabotropic glutamate 2/3 receptor agonists LY354740 and LY379268 selectively attenuate phencyclidine versus d-amphetamine motor behaviors in rats. J Pharmacol Exp Ther 291(1):161–170
65. Krystal JH et al (2005) Preliminary evidence of attenuation of the disruptive effects of the NMDA glutamate receptor antagonist, ketamine, on working memory by pretreatment with the group II metabotropic glutamate receptor agonist, LY354740, in healthy human subjects. Psychopharmacology (Berl) 179(1):303–309
66. Patil ST et al (2007) Activation of mGlu2/3 receptors as a new approach to treat schizophrenia: a randomized Phase 2 clinical trial. Nat Med 13(9):1102–1107
67. Galici R et al (2005) A selective allosteric potentiator of metabotropic glutamate (mGlu) 2 receptors has effects similar to an orthosteric mGlu2/3 receptor agonist in mouse models predictive of antipsychotic activity. J Pharmacol Exp Ther 315(3):1181–1187
68. Fell MJ et al (2008) Evidence for the role of metabotropic glutamate (mGlu)2 not mGlu3 receptors in the preclinical antipsychotic pharmacology of the mGlu2/3 receptor agonist (−)-(1R, 4S, 5S, 6S)-4-amino-2-sulfonylbicyclo[3.1.0]hexane-4, 6-dicarboxylic acid (LY404039). J Pharmacol Exp Ther 326(1):209–217
69. Woolley ML et al (2008) The mGlu2 but not the mGlu3 receptor mediates the actions of the mGluR2/3 agonist, LY379268, in mouse models predictive of antipsychotic activity. Psychopharmacology (Berl) 196(3):431–440
70. Johnson MP et al (2003) Discovery of allosteric potentiators for the metabotropic glutamate 2 receptor: synthesis and subtype selectivity of N-(4-(2-methoxyphenoxy)phenyl)-N-(2, 2, 2-trifluoroethylsulfonyl)pyrid-3-ylmethylamine. J Med Chem 46(15):3189–3192
71. Cube RV et al (2005) 3-(2-Ethoxy-4-{4-[3-hydroxy-2-methyl-4-(3-methylbutanoyl)phenoxy] butoxy}ph enyl)propanoic acid: a brain penetrant allosteric potentiator at the metabotropic glutamate receptor 2 (mGluR2). Bioorg Med Chem Lett 15(9):2389–2393
72. Pinkerton AB et al (2005) Allosteric potentiators of the metabotropic glutamate receptor 2 (mGlu2). Part 3: Identification and biological activity of indanone containing mGlu2 receptor potentiators. Bioorg Med Chem Lett 15(6):1565–1571
73. Galici R et al (2006) Biphenyl-indanone A, a positive allosteric modulator of the metabotropic glutamate receptor subtype 2, has antipsychotic- and anxiolytic-like effects in mice. J Pharmacol Exp Ther 318(1):173–185
74. Schaffhauser H et al (2003) Pharmacological characterization and identification of amino acids involved in the positive modulation of metabotropic glutamate receptor subtype 2. Mol Pharmacol 64(4):798–810

75. Rowe BA et al (2008) Transposition of three amino acids transforms the human metabotropic glutamate receptor (mGluR)-3-positive allosteric modulation site to mGluR2, and additional characterization of the mGluR2-positive allosteric modulation site. J Pharmacol Exp Ther 326 (1):240–251

76. Benneyworth MA et al (2007) A selective positive allosteric modulator of metabotropic glutamate receptor subtype 2 blocks a hallucinogenic drug model of psychosis. Mol Pharmacol 72(2):477–484

77. Poisik O et al (2005) Metabotropic glutamate receptor 2 modulates excitatory synaptic transmission in the rat globus pallidus. Neuropharmacology 49(Suppl 1):57–69

78. Govek SP et al (2005) Benzazoles as allosteric potentiators of metabotropic glutamate receptor 2 (mGluR2): efficacy in an animal model for schizophrenia. Bioorg Med Chem Lett 15(18):4068–4072

79. Aghajanian GK, Marek GJ (1997) Serotonin induces excitatory postsynaptic potentials in apical dendrites of neocortical pyramidal cells. Neuropharmacology 36(4–5):589–599

80. Marek GJ et al (2001) A major role for thalamocortical afferents in serotonergic hallucinogen receptor function in the rat neocortex. Neuroscience 105(2):379–392

81. Marek GJ, Wright RA, Schoepp DD (2006) 5-Hydroxytryptamine2A (5-HT2A) receptor regulation in rat prefrontal cortex: interaction of a phenethylamine hallucinogen and the metabotropic glutamate2/3 receptor agonist LY354740. Neurosci Lett 403(3):256–260

82. Gonzalez-Maeso J et al (2007) Hallucinogens recruit specific cortical 5-HT(2A) receptor-mediated signaling pathways to affect behavior. Neuron 53(3):439–452

83. Gonzalez-Maeso J et al (2008) Identification of a serotonin/glutamate receptor complex implicated in psychosis. Nature 452(7183):93–97

84. Molinaro G et al (2009) Activation of mGlu2/3 metabotropic glutamate receptors negatively regulates the stimulation of inositol phospholipid hydrolysis mediated by 5-hydroxytryptamine2A serotonin receptors in the frontal cortex of living mice. Mol Pharmacol 76(2):379–387

mGluR1 Negative Allosteric Modulators: An Alternative Metabotropic Approach for the Treatment of Schizophrenia

Hisashi Ohta, Hiroshi Kawamoto, and Gentaroh Suzuki

Abstract The potential utility of mGluR1 negative allosteric modulators (NAMs) for treatment of schizophrenia is based on the pharmacological effects of mGluR1 NAMs in animal models for schizophrenia. An mGluR1 NAM antagonized hyper-locomotion and the deficit in prepulse inhibition (PPI) in rodents caused by an indirect dopamine (DA) agonist, methamphetamine as well as by a noncompetitive N-methyl-D-aspartate (NMDA) receptor antagonist, ketamine. In addition, an mGluR1 NAM reversed reduced social interaction by dizocilpine (MK-801) in rats. The antipsychotic-like effects of mGluR1 NAMs are similar to those of the atypical antipsychotic, clozapine, but not of the typical antipsychotic, haloperidol, based on behavioral changes as well as distribution of c-fos expression after the treatment. The similarities and differences between mGluR1 NAMs and mGluR2/3 agonists are discussed.

Abbreviations

5-HT	Serotonin
AIDA	(RS)-1-aminoindan-1,5-dicarboxylic acid
BAY36-7620	(3aS,6aS)-6a-naphtalan-2-ylmethyl-5-methyliden-hexahydro-cyclopental[c]furan-1-on
CFMTI	2-cyclopropyl-5-[1-(2-fluoro-3-pyridinyl)-5-methyl-1H-1,2,3-triazol-4-yl]-2,3-dihydro-1H-isoindol-1-one
CNS	Central nervous system
CPCCOEt	7-(hydroxyimino)cyclopropa[b]chromen-1a-carboxylate ethyl ester
DA	Dopamine

H. Ohta (✉)
Research and Development Center, Sato Pharmaceutical Co., Ltd, 6-8-5 Higashi-ohi, Shinagawa, Tokyo 140-0011, Japan
e-mail: hisashi.ohta@sato-seiyaku.co.jp

P. Skolnick (ed.), *Glutamate-based Therapies for Psychiatric Disorders*,
Milestones in Drug Therapy, DOI 10.1007/978-3-0346-0241-9_7,
© Springer Basel AG 2010

DHPG	3,5-dihydroxyphenylglycine
dlSTR	dorsolateral striatum
EMQMCM	(3-ethyl-2-methyl-quinolin-6-yl)(4-methoxy-cyclohexyl)metha-none methansulfonate
FTIDC	4-[1-(2-fluoropyridin-3-yl)-5-methyl-1H-1,2,3-triazol-4-yl]-N-isopropyl-N-methyl-3,6-dihydropyridine-1(2H)-carboxamide
iGluR	ionotropic glutamate receptor
JNJ16259685	(3,4-dihydro-2H-pyrano[2,3]b quinolin-7-yl) (cis-4-methoxycy-clohexyl) methanone
LY354740	(1S,2S,5R,6S)-2-aminobicyclo[3.1.0] hexane-2,6-dicarboxylate monohydrate
LY367385	(+)-2-methyl-4-carboxyphenylglycine
LY379268	(1R,4R,5S,6R)-4-amino-2-oxabicyclo[3.1.0]hexane-4,6-dicarbox-ylic acid
LY456236	(4-methoxy-phenyl)-(6-methoxy-quinazolin-4-yl)-amine, HCl
MAP	Methamphetamine
mGluR	metabotropic glutamate receptor
MK-801	(5S,10R)-(+)-5-methyl-10,11-dihydro-5H-dibenzo[a,d]cyclohep-ten-5,10-imine maleate
MPEP	2-methyl-6-(phenylethynyl)pyridine
mPFC	medial prefrontal cortex
MTEP	3-[(2-methyl-1,3-thiazol-4-yl)ethynyl]pyridine
NAC	Nucleus accumbens
NAM	Negative allosteric modulator
NMDA	N-methyl-D-aspartate
PCP	Phencyclidine
PFC	Prefrontal cortex
PPI	Prepulse inhibition
VTA	Ventrotegmental area
YM-230888	N-cycloheptyl-6-({[(2R)-tetrahydrofuran-2-ylmethyl]amino} methyl)thieno[2,3-d]pyrimidin-4-amine

1 Introduction

Schizophrenia is a common mental disorder affecting 1% of the world population. The symptoms are classified into positive symptoms such as hallucinations and delusions, negative symptoms such as affective flattening and social withdrawal, and cognitive impairment to attention and working memory [1].

The discovery of chlorpromazine in the 1950s was a breakthrough in the pharmacological treatment of schizophrenia, and the success of chlorpromazine triggered the development of additional dopamine (DA) antagonists to treat the disease [2, 3]. For more than 20 years, various DA antagonists were introduced for the treatment of schizophrenia, despite a significant incidence of extrapyramidal syndrome and tardive dyskinesia with prolonged use [4, 5]. In the 1970s, the second generation or atypical antipsychotics, represented by clozapine, was introduced in the market. Clozapine remains one of the most effective agents for the treatment of schizophrenia [6]. Despite its effectiveness, the pharmacological profile of clozapine is promiscuous, since it interacts with various neurotransmitter receptors including DA D2 and serotonin (5-HT) 2a receptors [7]. Atypical antipsychotics available in the present market are still homologs of clozapine.

Abuse of phencyclidine (PCP, angel dust) and ketamine elicited psychosis in normal individuals and aggravated psychosis in schizophrenic patients [8, 9]. It has been demonstrated that both PCP and ketamine interact with N-methyl-D-aspartate (NMDA) receptors in the central nervous system (CNS), which strongly suggested an abnormality of glutamate transmission was involved in the altered brain function of schizophrenia. Moghaddam and Adams [10] demonstrated that activation of metabotropic glutamate receptor (mGluR)2/3 with LY354740 ameliorated some of behavioral and neurochemical changes caused by PCP in animals in 1998. Phase II clinical results with the mGluR2/3 agonist, LY404039, indicated that modulation of mGluR2/3 is effective in treating schizophrenia [11]. These results could potentially open a new era for development of new antipsychotics beyond the modulation of dopaminergic (or monoaminergic) neurotransmission (see chapter by Sheffler and Conn).

The mGluR family consists of eight receptor subtypes, which are divided into three groups based on sequence homology, pharmacological profiles, and signal transduction pathways [12, 13]. Group I mGluRs comprise mGluR1 and mGluR5, which are coupled with Gq to activate phospholipase C, leading to the release of intracellular calcium. Group II (mGluR2 and mGluR3) and Group III (mGluR4, mGluR6, mGluR7, and mGluR8) mGluRs are negatively coupled via Gi to adenylyl cyclase, resulting in the inhibition of cyclic adenosine monophosphate production and protein kinase A activity.

The mGluRs are expressed on neuronal and glial cells, with each receptor subtype exhibiting distinct spatial and temporal expression profiles in the brain [14], with the exception of mGluR6, present in the retina [15]. In neurons, group I mGluRs are mainly localized in somatodendritic domains and postsynaptically regulate neuronal excitability and synaptic transmission via several intracellular second messenger systems, whereas group II and III mGluRs are predominantly localized in axonal domains and axon terminals to presynaptically regulate neurotransmitter release [14].

mGluR2/3 localized in presynaptic terminals negatively regulate the release of glutamate, that is, they act as autoreceptors. mGluR2/3 agonists have shown anxiolytic, antinociceptive, and antipsychotic effects in animal models. Thus, modulation of glutamatergic neurotransmission has been considered as a potential therapeutic target for these CNS disorders [16–19]. Alteration of glutamatergic neurotransmission can be achieved not only by presynaptic regulation of glutamate release with mGluR2/3, but also by postsynaptic modulation with iGluRs, mGluR1, and/or mGluR5. Studies with NMDA receptor, non-NMDA receptor, and mGluR5 antagonists suggest the pharmacological effects of mGluR2/3 agonists via reduction of glutamatergic neurotransmission cannot be fully explained by postsynaptic block-ade of iGluRs and mGluR5 [20, 21]. In contrast to studying the pharmacological effects of iGluRs and mGluR5 antagonists, studies to explore pharmacological effects of mGluR1 antagonists have been limited, due to the lack of selective antagonists, until the recent discovery of mGluR1 negative allosteric modulators (NAMs).

2 Development of mGluR1 Negative Allosteric Modulators (NAMs)

Amino acid-derived ligands such as 3,5-dihydroxyphenylglycine (DHPG) and AIDA were used to explore function of mGluR1 in earlier studies. Although the potency and selectivity of amino acid-derived agents were limited, intra-nucleus accumbens (NAC) injection of DHPG impaired prepulse inhibition (PPI) [22]. Demonstration of pharmacological effects by blockade of mGluR1 has been hampered by the lack of appropriate antagonists until the recent development of negative allosteric modula-tors, including JNJ16259685, BAY 36-7620, YM-230888, and EMQMCM. A limited study examining the antipsychotic activities of BAY 36-7620 at 10 mg/kg reported efficacy in suppression of stereotypic behaviors induced by MK-801, but not by amphetamine or apomorphine [23], and PPI disruption by MK-801, PCP, and apo-morphine [13]. However, in ex vivo occupancy studies, BAY 36-7620 at 10 mg/kg only occupied about 30% of cerebellar and thalamic mGluR1 [24], suggesting that BAY 36-7620 at doses in these reports might be insufficient to fully block mGluR1. Although JNJ16259685 demonstrated nearly full mGluR1 occupancy at 0.16 mg/kg, it has not been tested on these antipsychotic activities [25]. A newly discovered 4-[1-(2-fluoropyridin-3-yl)-5-methyl-1H-1,2,3-triazol-4-yl]-N-isopropyl-N-methyl-3, 6-dihydropyridine-1(2H)-carboxamide (FTIDC) and2-cyclopropyl-5-[1-(2-fluoro-3-pyridinyl)-5-methyl-1 H-1,2,3-triazol-4-yl]-2,3-dihydro-1H-isoindol-1-one (CFMTI) (Fig. 1) are potent selective mGluR1 NAMs [26–28]. It will be important to clarify the relationship between receptor occupancy and doses of mGluR1 NAMs to examine pharmacological effects of mGluR1 NAMs. In the case of FTIDC and CFMTI, the relationship between receptor occupancy and doses has been documented (Table 1), together with their in vivo antagonistic activity [26–28]. In vivo antagonistic activity

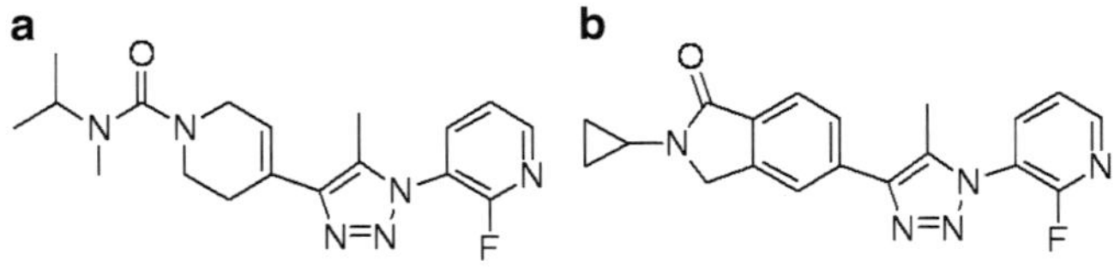

Fig. 1   Chemical structures of (**a**) FTIDC [4-[1-(2-fluoropyridin-3-yl)-5-methyl-1 H-1,2,3-triazol-4-yl]-N-isopropyl-N-methyl-3,6-dihydropyridine-1(2H)-carboxamide] and (**b**) CFMTI [2-cyclopro-pyl-5-[1-(2-fluoro-3-pyridinyl)-5-methyl-1H-1,2,3-triazol-4-yl]-2,3- dihydro-1H-isoindol-1-one]

Table 1 Ex vivo mGluR1 occupancy in striatum and cerebellum, after administration of FTIDC and CFMTI in mice

Dose mg/kg	mGluR1 occupancy (%)		n
	Striatum	Cerebellum	
FTIDC			
10	46 ± 3.6	18 ± 6.0	3
30	75 ± 2.2	60 ± 3.5	3
100	92 ± 3.2	88 ± 5.2	3
CFMTI			
3	73 ± 7	70 ± 8	3
10	87 ± 1	82 ± 2	3
30	94 ± 1	89 ± 2	3

Receptor occupancy and drug concentration were expressed as means ± SEM.

Ex vivo occupancy was measured with ^{3}H-labeled FTIDC 30 min and 60 min after intraperitoneal administration of FTIDC and oral administration of CFMTI, respectively.

was demonstrated against DHPG-induced face washing behavior, which was selectively mediated by mGluR1 but not mGluR5 in mice [30].

3   Effects on DA-Dependent Behavior in Animal Models

Behavioral abnormalities elicited by DA-direct/indirect agonists have been used to develop antipsychotics in preclinical studies. When the effects of mGluR1 NAMs were tested on behavioral changes elicited by methamphetamine (MAP) in rodents [27, 28], both FTIDC and CFMTI decreased MAP-induced hyperlocomotion in a dose-dependent manner (Fig. 2) without affecting basal spontaneous locomotion. An mGluR2/3 agonist, LY379268 inhibited hyperlocomotion induced by MAP with reduction of spontaneous locomotor activity [27]. An mGluR5 NAM, MPEP, attenuated MAP-induced hyperlocomotion only at a high dose which caused hypolocomotion by itself [27]. In this model, both typical and atypical antipsychotics attenuated the hyperlocomotion with reduction of spontaneous locomotor activity [27, 28]. Intraventrotegmental area (VTA) injection of DHPG raised extracellular DA in the medial prefrontal cortex (mPFC) and intra-VTA coadministration of CPCCOEt or MPEP reversed increases in DA by DHPG [31]. Intra-NAC injection of an mGluR2/3 agonist reduced intra-NAC extracellular DA [32]. Although the exact mechanisms for mGluRs regulating release of DA might be different in specific brain areas, the neuronal activity of dopaminergic system is regulated by mGluRs to some extent.

Schizophrenic patients exhibit impairments in sensorimotor gating [33]. These impairments likely reflect disruption of the information filtering mechanism, which contributes to overloading of irrelevant stimuli in cortical regions and subsequently causes confusion, hallucinations, and attentional/cognitive deficits [34]. One way to assess impaired sensorimotor gating is to measure prepulse inhibition (PPI) of the startle reflex, in which the involuntary startle reflex is reduced when a startling stimulus is preceded by a weak stimulus. Consistent with DA-, glutamate-, and

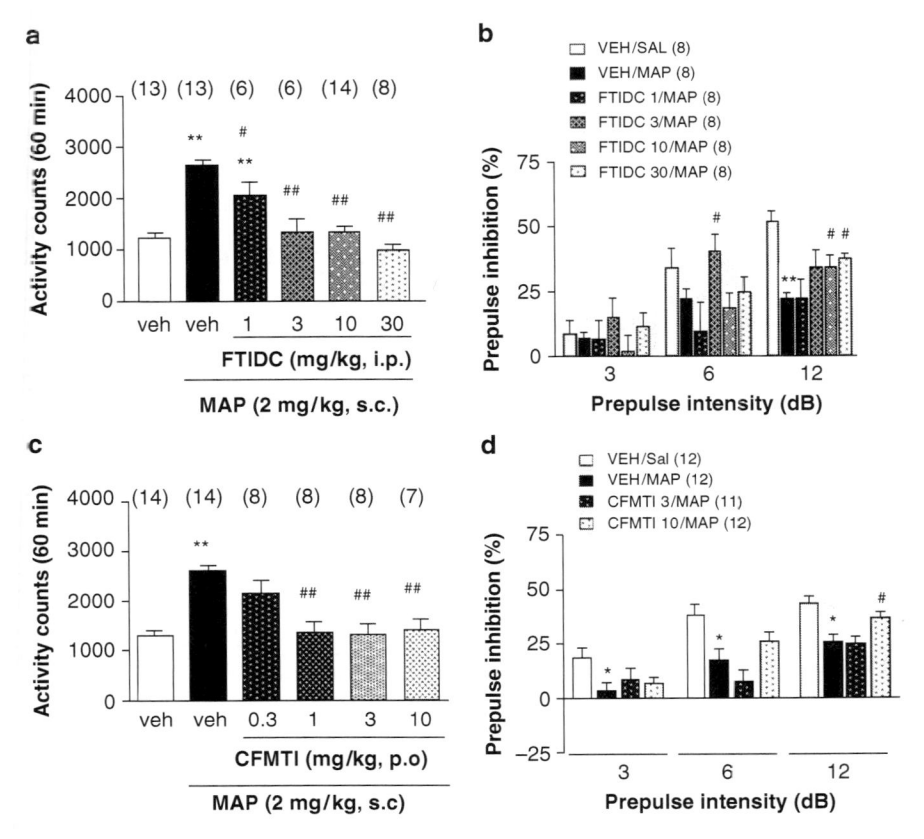

Fig. 2 Effects of mGluR1 NAMs on methamphetamine-induced hyperlocomotion in mice (**a, c**) and disruption of PPI in rats (**b, d**). (**a, b**) Effects of FTIDC and (**c, d**) effects of CFMTI. Data are presented as means ± SEM. *MAP* methamphetamine; *veh* vehicle The number of animals used in each group is indicated in parenthesis. *$p < 0.05$ and **$p < 0.01$ vs. vehicle-treated group, and #$p < 0.05$ and ##$p < 0.01$ vs. methamphetamine-treated group (one-way analysis of variance followed by Dunnett's test). Doses are in mg/kg

5-HT-based hypotheses of schizophrenia, PPI in rats can be disrupted by direct or indirect dopamine agonists, NMDA receptor antagonists, and 5-HT$_{2A}$ receptor agonists. Reversal of these pharmacologically induced deficits in PPI is considered as one of the indices for potential antipsychotic properties. The role of mGluR1 in sensorimotor gating has been studied with mGluR1 deficient mice [35]. Although mGluR1 deficient mice exhibited a deficit of PPI, it should be noted that mGluR1 deficient mice exhibited severe cerebellar ataxia [36].

MAP-induced deficits of PPI were significantly reversed by both FTIDC and CFMTI at doses which attenuated hyperlocomotion induced by MAP (Fig. 2). BAY 36-7620 did not alter a deficit of PPI elicited by apomorphine [23], although this dose of BAY 36-7620 might not be sufficient enough to block mGluR1 due to low receptor occupancy [24]. In contrast to mGluR1 NAMs, an mGluR5 NAM,

MPEP exaggerated MAP-induced disruption of PPI without affecting spontaneous PPI, consistent with findings that a selective mGluR5 positive allosteric modulator (PAM) antagonized amphetamine-induced PPI deficits [37], while LY379268 affected neither impaired PPI nor spontaneous PPI [27]. Postsynaptic blockade of mGluR1 and mGluR5 could produce opposite pharmacological effects. Presynaptic inhibition of glutamatergic neuronal activities by mGluR2/3 activation could result in summation of postsynaptic blockade of both mGluR1 and mGluR5, at least in the dopamine-dependent neuronal system [27], which might be one of reasons why an mGluR2/3 agonist did not reverse MAP-induced disruption of PPI despite antagonizing hyperlocomotion elicited by MAP. Blockade of mGluR1 ameliorated DA-mediated behavioral changes on both locomotor and PPI in rodents, which are similar to the results obtained by blockade of DA receptors, while mGluR2/3 agonists preferentially inhibited DA-mediated locomotor changes but not PPI.

4 Effects on Glutamate (NMDA)-Dependent Behavior in Animal Models

The glutamate hypothesis in schizophrenia is rooted in the clinical observations that PCP and ketamine are known to elicit psychosis in human and exaggerate symptoms of schizophrenia [8, 9].

mGluR1 NAMs, FTIDC, and CFMTI inhibit ketamine-induced hyperlocomotion (Fig. 3). BAY 36-7620 at 10 mg/kg suppressed stereotypic behaviors induced by MK-801. Both the typical antipsychotic haloperidol and the atypical clozapine decreased ketamine-induced hyperlocomotion [28].

The effects of mGluR1 antagonists on PPI disrupted by NMDA antagonists have not been fully elucidated. Spooren et al. [13] mentioned no effects of BAY 36-7620 on MK-801-induced PPI disruption. EMQMCM, studied only at 4 mg/kg, was shown to be ineffective in MK-801-induced PPI disruption [38]. CFMTI reversed ketamine-induced deficits in PPI. Clozapine antagonized against PPI disruption induced by ketamine (Fig. 3).

MK-801-induced deficits in social interaction may be an animal model for negative symptoms of schizophrenia, particularly social withdrawal. In this test, CFMTI and clozapine improved MK-801-induced social withdrawal without affecting locomotor activity, indicating that positive effects of these drugs on social withdrawal are not due to apparent behavioral activation. Haloperidol did not reverse MK-801-induced social withdrawal (Fig. 4).

Because it has been demonstrated that acute administration of NMDA receptor antagonists such as PCP, ketamine, and MK-801 facilitate release of glutamate in the prefrontal cortex in rodents [39], increased extracellular levels of glutamate would be involved in NMDA receptor antagonist-induced behavioral alterations, such as PPI disruption and/or deficits in social interaction. Although enhanced glutamate release, hyperlocomotion and stereotypy elicited by PCP were blocked

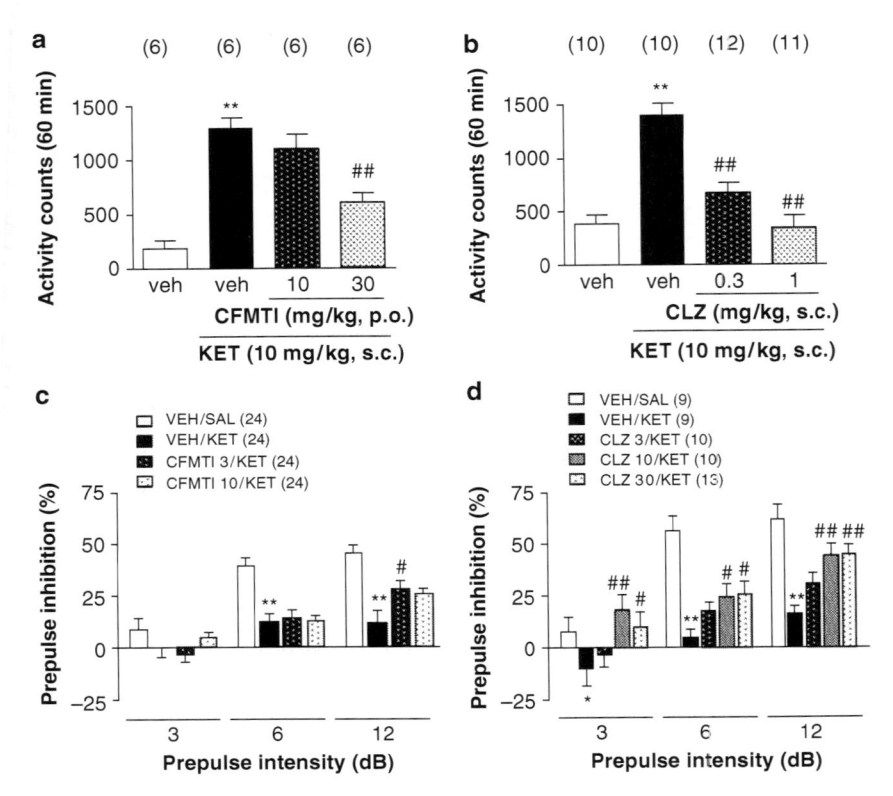

Fig. 3 Effects of mGluR1 NAM and clozapine on ketamine-induced hyperlocomotion in mice (**a, b**) and disruption of PPI in rats (**c, d**). (**a, c**) Effects of CFMTI. (**b, d**) effects of clozapine. Data are presented as means ± SEM. The number of animals used in each group is indicated in parenthesis. Doses are in mg/kg. *VEH* vehicle; *SAL* saline; *KET* ketamine (5 mg/kg, s.c.); *CLZ* clozapine. $*p < 0.05$ and $**p < 0.01$ vs. vehicle-treated group, $\#p < 0.05$ and $\#\#p < 0.01$ vs. ketamine-treated group (one-way analysis of variance followed by Dunnett's test)

by the mGluR2/3 agonist LY354740 [10], mGluR2/3 agonists had only limited effects on PCP-induced PPI disruption [40]. At present, it is not readily explained why mGluR2/3 agonists only reversed hyperlocomotion and stereotypy elicited by PCP but not PPI disruption. These results suggest that presynaptic control of the glutamatergic nervous system is apparently less significant in PCP-induced PPI disruption. Blockade of postsynaptic mGluR5 by MPEP or MTEP was demonstrated to exacerbate NMDA receptor antagonist-induced PPI disruption [37, 38]. Activation of presynaptic mGluR5 appeared to release glutamate [41], which may further complicate behavioral outcomes. Melendez et al. [42] reported that DHPG administered by reverse microdialysis into the mPFC elicited glutamate release, which was blocked by coapplication of a competitive mGluR1 antagonist, AIDA, and a group I mGluRs antagonist, LY367385 with DHPG, suggesting that mGluR1 in the mPFC may play a role in regulating glutamate release. Blockade of

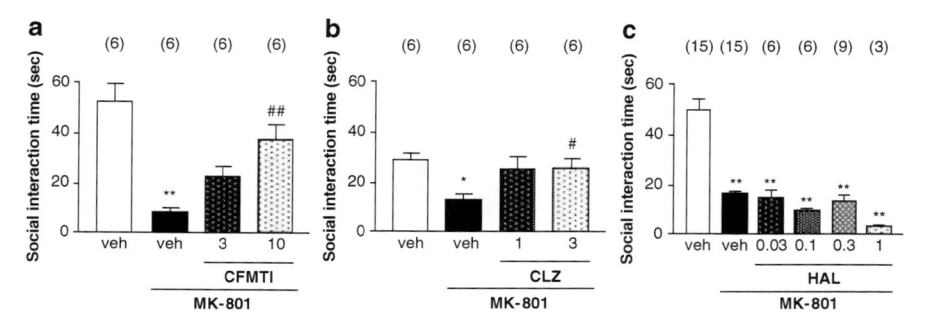

Fig. 4 Effects on MK-801-induced deficits in social interaction. (**a**) Effects of CFMTI. (**b**) Effects of clozapine. (**c**) Effects of haloperidol. Data are presented as mean ± SEM. Doses are in mg/kg. *Veh* vehicle; *HAL* haloperidol; *CLZ* clozapine. Experimental numbers of each group are indicated in parentheses. $*p < 0.05$ and $**p < 0.01$ vs. vehicle-treated group, and $\#p < 0.05$ and $\#\#p < 0.01$ vs. MK-801-treated group (one-way analysis of variance followed by Dunnett's test)

postsynaptic mGluR1 might suppress excess glutamate neurotransmission elicited by NMDA receptor antagonists in the mPFC. Increases in L-glutamate levels in the mPFC correlated with hyperlocomotion and cognitive impairment elicited by PCP in rats [10]. The PCP-induced hyperglutamatergic state in the PFC could result from inhibition of GABAergic interneurons in the PFC caused by blockade of NMDA receptor-dependent excitatory input and subsequent disinhibition of glutamatergic neurons in the PFC [43]. mGluR1 NAM might counteract such activation of glutamatergic activity in the PFC and output from the PFC.

5 Immunohistochemical Analysis of c-fos Expression in Rats

Since expression of the immediate early gene c-fos is increased by neuronal activation, induction of c-fos in neurons is a useful marker to map changes in neuronal activity. Antipsychotics are known to induce c-fos in various brain regions. Among them, expression of c-fos in the NAC might be related, at least to some extent, to the antipsychotic action of typical and atypical antipsychotics. In addition to c-fos induction in the NAC, the atypical antipsychotic, clozapine, is known to induce c-fos in the mPFC, whereas typical antipsychotics do not. Fos expression in the mPFC could be associated with potency for the treatment of negative symptoms in schizophrenia [44]. On the other hand, induction of c-fos in the dorsolateral striatum (dlSTR) might be a response to extrapyramidal side effects of antipsychotics. Therefore, profiling the fos expression pattern could aid in the classification of novel compounds and predict their therapeutic utility [45]. The mGluR1 NAM, CFMTI, significantly increased numbers of c-fos positive cells in the NAC as well as in the mPFC, but not in the dlSTR (Fig. 5). These observations with CFMTI were similar to changes caused by atypical antipsychotics clozapine. CFMTI did not cause induction of c-fos in the dlSTR might be related to noncataleptogenic activity of

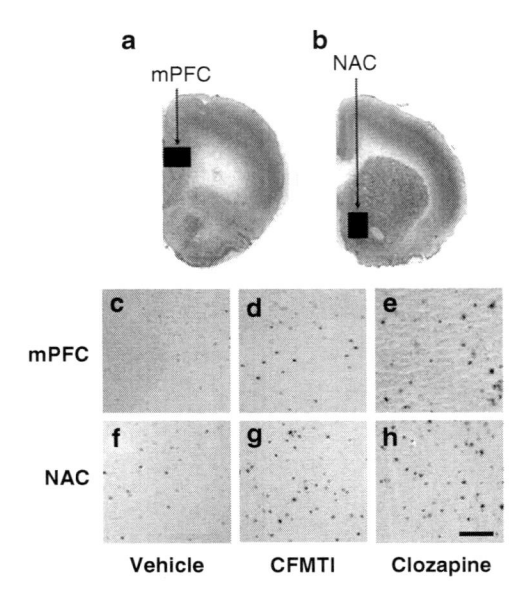

Fig. 5 Representative sections used for counting of fos-positive neurons (a and b) and photomicrographs of fos immunoreactivity (**c–h**). (**a**) Location of medial prefrontal cortex, corresponding to an AP position ~2.7 mm from bregma. (**b**) Location of nucleus accumbens, corresponding to an AP position ~2.2 mm from bregma. (**c–e**) c-Fos induction in the nucleus accumbens. (**f–h**) c-Fos induction in the medial prefrontal cortex. (**c** and **f**) Vehicle treatment. (**d**, **g**) CFMTI (30 mg/kg, p.o.) treatment. (**e**, **h**) Clozapine (30 mg/kg, i.p.) treatment. *NAC* nucleus accumbens; *mPFC* medial prefrontal cortex. Scale bar = 100 μm

CFMTI. Although mGluR2/3 agonists have been studied in various behavioral and neurochemical aspects in animal models of schizophrenia, the specific c-fos expression in brain areas was not intensively studied. Taken together, the behavioral effects and distribution of c-fos expression by treatment of mGluR1 NAM, mGluR1 NAM and clozapine might either share the same neuronal circuits or at least neuronal circuits altered by these two agents might be overlapped to some extent.

6 Potential Side Effects by mGluR1 NAMs

6.1 Motor Coordination

The striking phenotype of mGluR1 deficient mouse is severe cerebellar ataxia. An important anatomical abnormality in the cerebellum in mGluR1 knock-out mice is persistent multiple climbing fiber innervations of Purkinje cells without apparent defect in parallel fiber-Purkinje cell synaptogenesis [36]. Cerebellar ataxia in mGluR1 knock-out mice is not thought to be entirely the consequence of persistent multiple innervation of climbing fibers on Purkinje cells, because

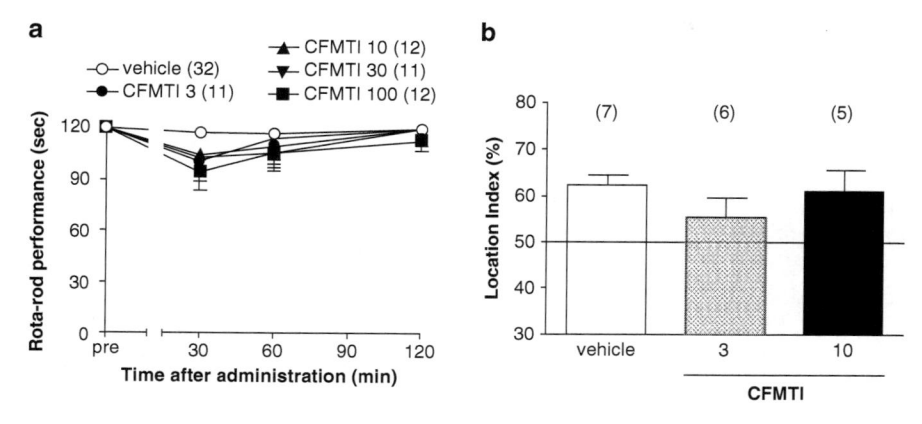

Fig. 6 Effects of CFMTI on motor coordination with a rota-rod test in mice (**a**) and on spatial memory with an object location test (**b**). Data are presented as mean $\pm$ SEM. Doses are in mg/kg. Some *upper* or *lower* error bars are omitted to simplify the figure. Experimental numbers of each group are indicated in parentheses. $*p < 0.05$ and $**p < 0.01$ vs. vehicle-treated group by Dunnett's test

mGluR1 autoantibodies caused motor coordination deficits in patients with Hodgkin's disease who had normal maturation of the cerebellar cortex [46]. Thus, one of the concerns about the potential side effects with mGluR1 NAMs was cerebellar motor incoordination. Both CFMTI and FTIDC at doses produced antipsychotic-like effects in rodents did not cause motor incoordination evaluated by a rotarod test (Fig. 6). Pharmacological blockade of mGluR1 did not cause severe motor incoordination, unlike mGluR1 deficient mice.

6.2 Cognitive Behaviors

Several studies have demonstrated the involvement of mGluR1 in spatial and associative learning. JNJ16259685 was demonstrated to produce more than 80% mGluR1 occupancy in rat brain at and above 0.63 mg/kg [25]. At these doses, impairment of spatial acquisition in the Morris water maze was observed in mice [47]. However, CFMTI did not impair spatial memory task in the object location test [28] at a dose which demonstrated antipsychotic activities in animal models (10 mg/kg). The object location test did not require any reinforcement to perform the task unlike water maze task [48]. Involvement of mGluR1 in cognitive behaviors may depend on both tasks and experimental conditions. A hallmark of schizophrenia is cognitive impairment in addition to positive and negative symptoms, and treatments with antipsychotics improve cognitive dysfunction [49]. However, there are not many studies available to evaluate cognitive performance after antipsychotic treatment in animal models. It is not yet known if mGluR1 antagonists improve cognitive deficits caused by NMDA receptor antagonists.

7 Conclusion

mGluR1 NAMs exhibit antipsychotic-like activity in both DA-dependent and NMDA-dependent animal models. MK-801 induced reduction in rodent social interaction was reversed by mGluR NAM and clozapine, but not by haloperidol. In general, the antipsychotic-like effects of mGluR1 NAMs are similar to those of atypical antipsychotic, clozapine, but not of haloperidol. The similarity between mGluR1 NAM and clozapine was observed not only in behavioral effects but also in expression pattern of c-fos in the brain. Most behavioral changes indicative of the antipsychotic activities of mGluR1 NAMs are similar to those produced by mGluR2/3 agonists, except that the deficit in PPI by both DA agonists and NMDA receptor antagonists is only reversed by mGluR1 NAM. An mGluR2/3 agonist has demonstrated efficacy in human. Thus, it would be interesting to see how the differences in preclinical models might reflect in clinical studies with mGluR1 NAMs. We hope the effort to develop bioavailable and safe mGluR1 NAMs for clinical trials is continued.

Acknowledgments We thank Satoru Itoh, Hirohiko Hikichi, Shunsuke Maehara, Toshifumi Kimura, and Akio Satow who conducted studies with FTIDC and CFMTI.

References

1. Andreasen NC (2000) Schizophrenia: the fundamental questions. Brain Res Rev 31:106–112
2. Laborit H (1949) Sur l'utilization de certain agents pharmacodynamiques a action neuro-vegetative en periode per- and post-operatioire. Acta Chir Belg 87:485–492
3. Carlsson A, Lindquist M (1963) Effect of chlorpromazine and haloperidol on the formation of 3-methoxytyramine in mouse brain. Acta Pharmacol Toxicol 20:140–144
4. Sovner R, DiMascio A (1978) Extrapyramdal syndrome and other neurological side effects of psychotropic drugs. In: Lipton MA, DiMascio A, Killam KF (eds) Psychopharmacology: a generation of progress. Raven Press, New York, pp 1021–1032
5. Baldessarini RJ, Tarsy D (1978) Tardive dyskinesia. In: Lipton MA, DiMascio A, Killam KF (eds) Psychopharmacology: a generation of progress. Raven Press, New York, pp 993–1004
6. Kane J, Honigfeld G, Singer J, Meltzer H (1988) Clozapine for the treatment-resistant schizophrenic A double-blind comparison with chlorpromazine. Arch Gen Psychiatry 45:789–796
7. Sur C, Mallorga PJ, Wittmann M, Jacobson MA, Pascarella D, Williams JB, Brandish PE, Pettibone DJ, Scolnick EM, Conn PJ (2003) N-desmethylclozapine, an allosteric agonist at muscarinic 1 receptor, potentiates N-methyl-D-aspartate receptor activity. Proc Natl Acad Sci USA 100:13674–13679
8. Javitt DC, Zukin SR (1991) Recent advance in the phencyclidine model of schizophrenia. Am J Psychiatry 148:1301–1308
9. Lahti AC, Koffel B, LaPorte D, Tamminga CA (1995) Subanesthetic doses of ketamine stimulate psychosis in schizophrenia. Neuropsychopharmacology 13:9–19
10. Moghaddam B, Adams BW (1998) Reversal of phencyclidine effects by a group II metabotropic glutamate receptor agonist in rats. Science 281:1349–1352
11. Patil ST, Zhang L, Martenyi F, Lowe SL, Jackson KA, Andreev BV, Avedisova AS, Bardenstein LM, Gurovich IY, Morozova MA et al (2007) Activation of mGlu2/3 receptors

as a new approach to treat schizophrenia: a randomized phase 2 clinical trial. Nat Med 13:1102–1107

12. De Blasi A, Conn PJ, Pin JP, Nicoletti F (2001) Molecular determinants of metabotropic glutamate receptor signaling. Trends Pharmacol Sci 22:114–120

13. Spooren W, Ballard T, Gasparini F, Amalric M, Mutel V, Schreiber R (2003) Insight into the function of group I and group II metabotropic glutamate (mGlu) receptors: behavioral characterization and implications for the treatment of CNS disorders. Behav Pharmacol 14:257–277

14. Shigemoto R, Nakanishi S, Mizuno N (1992) Distribution of the mRNA for a metabotropic glutamate receptor (mGluR1) in the central nervous system: an in situ hybridization study in adult and developing rat. J Comp Neurol 322:121–135

15. Nakajima Y, Iwakabe H, Akazawa C, Nawa H, Shigemoto R, Mizuno N, Nakanishi S (1993) Molecular characterization of a novel retinal metabotropic glutamate receptor mGluR6 with a high agonist selectivity for L-2-amino-4-phosphonobutyrate. J Biol Chem 268: 11868–11873

16. Kłodzińska A, Chojnacka-Wójcik E, Pałucha A, Brański P, Popik P, Pilc A (1999) Potential anti-anxiety, anti-addictive effects of LY 354740, a selective group II glutamate metabotropic receptors agonist in animal models. Neuropharmacology 38:1831–1839

17. Cartmell J, Monn JA, Schoepp DD (2000) The mGlu(2/3) receptor agonist LY379268 selectively blocks amphetamine ambulations and rearing. Eur J Pharmacol 400:221–224

18. Grillon C, Cordova J, Levine LR, Morgan CA III (2003) Anxiolytic effects of a novel group II metabotropic glutamate receptor agonist (LY354740) in the fear-potentiated startle paradigm in humans. Psychopharmacology (Berl) 168:446–454

19. Simmons RM, Webster AA, Kalra AB, Iyengar S (2002) Group II mGluR receptor agonists are effective in persistent and neuropathic pain models in rats. Pharmacol Biochem Behav 73:419–427

20. Fundytus ME (2001) Glutamate receptors and nociception. CNS Drugs 15:29–58

21. Javitt DC (2004) Glutamate as a therapeutic target in psychiatric disorders. Mol Psychiatry 9:984–997

22. Grauer SM, Marquis KL (1999) Intracerebral administration of metabotropic glutamate receptor agonists disrupts prepulse inhibition of acoustic startle in Sprague-Dawley rats. Psychopharmacology (Berl) 141:405–412

23. De Vry J, Horváth E, Schreiber R (2001) Neuroprotective and behavioral effects of the selective metabotropic glutamate mGlu$_1$ receptor antagonist BAY 36-7620. Eur J Pharmacol 428:203–214

24. Lavreysen H, Pereira SN, Leysen JE, Langlois X, Lesage ASJ (2004) Metabotropic glutamate 1 receptor distribution and occupancy in the rat brain: a quantitative autoradiographic study using [3H]R214127. Neuropharmacology 46:609–619

25. Lavreysen H, Wouters R, Bischoff F, Nóbrega Pereira S, Langlois X, Blokland S, Somers M, Dillen L, Lesage ASJ (2004) JNJ16259685, a highly potent, selective and systemically active mGlu1 receptor antagonist. Neuropharmacology 47:961–972

26. Suzuki G, Kimura T, Satow A, Kaneko N, Fukuda J, Hikichi H, Sakai N, Maehara S, Kawagoe-Takaki H, Hata M et al (2007) Pharmacological characterization of a new, orally active and potent allosteric metabotropic glutamate receptor 1 antagonist, 4-[1-(2-Fluoropyridin-3-yl)-5-methyl-1H–1, 2, 3-triazol-4-yl]-N-isopropyl-N-methyl-3, 6- dihydropyridine-1 (2H)-carboxamide (FTIDC). J Pharmacol Exp Ther 321:1144–1153

27. Satow A, Maehara S, Ise S, Hikichi H, Fukushima M, Suzuki G, Kimura T, Tanaka T, Ito S, Kawamoto H et al (2008) Pharmacological effects of the metabotropic glutamate receptor 1 antagonist compared with those of the metabotropic glutamate receptor 5 antagonist and metabotropic glutamate receptor 2/3 agonist in rodents: detailed investigations with a selective allosteric mGluR1 antagonist, FTIDC, (4-[1-(2-fluoropyridine-3-yl)-5-methyl-1H-1,2,3-triazol-4-yl]-N-isopropyl-N-methyl-3,6-dihydropyridine-1(2H)-carboxamide). J Pharmacol Exp Ther 326:577–586

28. Satow A, Suzuki G, Maehara S, Hikichi H, Murai T, Murai T, Kawagoe-Takaki H, Hata M, Ito S, Ozaki S et al (2009) Unique antipsychotic activities of the selective metabotropic glutamate receptor 1 allosteric antagonist 2-cyclopropyl-5-[1-(2-fluoro-3-pyridinyl)-5-methyl-1H–1, 2, 3-triazol-4-yl]-2, 3-dihydro-1H-isoindol-1-one. J Pharmacol Exp Ther 330:179–190

29. Suzuki G, Kawagoe-Takaki H, Inoue T, Kimura T, Hikichi H, Murai T, Satow A, Hata M, Maehara S, Ito S et al (2009) Correlation of receptor occupancy of metabotropic glutamate receptor subtype 1 (mGluR1) in mouse brain with in vivo activity of allosteric mGluR1 antagonists. J Pharmacol Sci 110:315–325

30. Hikichi H, Iwahori Y, Murai T, Maehara S, Satow A, Ohta H (2008) Face-washing behavior induced by the group I metabotropic glutamate receptor agonist (S)-3, 5-DHPG in mice is mediated by mGlu1 receptor. Eur J Pharmacol 586:212–216

31. Renoldi G, Calcagno E, Borsini F, Invernizzi RW (2006) Stimulation of group I mGlu receptors in the ventrotegmental area enhances extracellular dopamine in the rat medial prefrontal cortex. J Neurochem 100:1658–1666

32. Greenslade RG, Mitchell SN (2004) Selective action of (-)-2-oxa-4-aminobicyclo[3.1.0] hexane-4, 6-dicarboxylate (LY379268), a group II metabotropic glutamate receptor agonist, on basal and phencyclidine-induced dopamine release in the nucleus accumbens shell. Neuropharmacology 47:1–8

33. Braff DL, Grillon C, Geyer MA (1992) Gating and habituation of the startle reflex in schizophrenic patients. Arch Gen Psychiatry 49:206–215

34. Carlsson A, Hansson LO, Waters N, Carlsson ML (1997) Neurotransmitter aberrations in schizophrenia: new perspectives and therapeutic implications. Life Sci 61:75–94

35. Brody SA, Conquet F, Geyer MA (2003) Disruption of prepulse inhibition in mice lacking mGluR1. Eur J Neurosci 18:3361–3366

36. Aiba A, Kano M, Chen C, Stanton ME, Fox GD, Herrup K, Zwingman TA, Tonegawa S (1994) Deficient cerebellar long-term depression and impaired motor learning in mGluR1 mutant mice. Cell 79:377–388

37. Kinney GG, Burno M, Campbell UC, Hernandez LM, Rodriguez D, Bristow LJ, Conn PJ (2003) Metabotropic glutamate receptor subtype 5 receptors modulate locomotor activity and sensorimotor gating in rodents. J Pharmacol Exp Ther 306:116–123

38. Pietraszek M, Gravius A, Schäfer D, Weil T, Trifanova D, Danysz W (2005) mGluR5, but not mGluR1, antagonist modifies MK-801-induced locomotor activity and deficit of prepulse inhibition. Neuropharmacology 49:73–85

39. Moghaddam B, Adams B, Verma A, Daly D (1997) Activation of glutamatergic neurotransmission by ketamine: a novel step in the pathway from NMDA receptor blockade to dopaminergic and cognitive disruptions associated with the prefrontal cortex. J Neurosci 17:2921–2922

40. Schreiber R, Lowe D, Voerste A, De Vry J (2000) LY354740 affects startle responding but not sensorimotor gating or discriminative effects of phencyclidine. Eur J Pharmacol 388: R3–R4

41. Musante V, Neri E, Feligioni M, Puliti A, Pedrazzi M, Conti V, Usai C, Diaspro A, Ravazzolo R, Henley JM, Battaglia G, Pittaluga A (2008) Presynaptic mGlu1 and mGlu5 autoreceptors facilitate glutamate exocytosis from mouse cortical nerve endings. Neuropharmacology 55:474–482

42. Melendez RI, Vuthiganon J, Kalivas PW (2005) Regulation of extracellular glutamate in the prefrontal cortex: focus on the cystine glutamate exchanger and group I metabotropic glutamate receptors. J Pharmacol Exp Ther 314:139–147

43. Paz RD, Tardito S, Atzori M, Tseng KY (2008) Glutamatergic dysfunction in schizophrenia: from basic neuroscience to clinical psychopharmacology. Eur Neuropsychopharmacol 18:773–786

44. Robertson GS, Matsumura H, Fibiger HC (1994) Induction patterns of Fos-like immunoreactivity in the forebrain as predictors of atypical antipsychotic activity. J Pharmacol Exp Ther 271:1058–1066

45. Sumner BE, Cruise LA, Slattery DA, Hill DR, Shahid M, Henry B (2004) Testing the validity of c-fos expression profiling to aid the therapeutic classification of psychoactive drugs. Psychopharmacology (Berl) 171:306–321
46. Sillevis Smitt P, Kinoshita A, De Leeuw B, Moll W, Coesmans M, Jaarsma D, Henzen-Logmans S, Vecht C, De Zeeuw C et al (2000) Paraneoplastic cerebellar ataxia due to autoantibodies against a glutamate receptor. N Engl J Med 342:21–27
47. Steckler T, Oliveira AF, Van Dyck C, Van Craenendonck H, Mateus AM, Langlois X, Lesage ASJ, Prickaerts J (2005) Metabotropic glutamate receptor 1 blockade impairs acquisition and retention in a spatial Water maze task. Behav Brain Res 164:52–60
48. Murai T, Okuda S, Tanaka T, Ohta H (2007) Characteristics of object location memory in mice: behavioral and pharmacological studies. Physiol Behav 90:116–124
49. Meltzer HY, McGurk SR (1999) The effects of clozapine, risperidone, and olanzapine on cognitive function in schizophrenia. Schizophr Bull 25:233–255

Metabotropic Glutamate Receptors as Targets for the Treatment of Drug and Alcohol Dependence

Svetlana Semenova and Athina Markou

Abstract Experimental studies in laboratory animals discussed in this chapter demonstrate the involvement of various metabotropic glutamate receptors (mGluRs) in different aspects of drug and alcohol dependence. Postsynaptic mGluR5 antagonists and inhibitory presynaptic mGluR2/3 agonists decreased drug self-administration and attenuated reinstatement of drug-seeking behavior by reducing the increases in glutamate transmission induced by drugs of abuse or the presentation of stimuli previously associated with the drug effects or availability. These findings suggest that medications decreasing glutamatergic transmission may reduce the reinforcing and motivational properties of drugs of abuse and prevent relapse to drug taking in humans. mGluR2/3 antagonists may be useful in the treatment of depressive-like affective symptoms of drug withdrawal by reversing the hypothesized decrease in glutamate transmission that occurs during psychostimulant, but not opiate, withdrawal. The potential of these compounds as medications for drug and alcohol dependence remains to be evaluated in humans.

1 Introduction

Drug addiction is a chronically relapsing disorder. Dependence on drugs of abuse and alcohol is often defined by: (1) the persistence of drug-taking behavior despite adverse consequences, (2) the emergence of withdrawal symptoms upon abrupt cessation of drug administration (e.g., dysphoria, anhedonia, anxiety, irritability), and (3) relapse to drug taking even after a period of extended abstinence after the

A. Markou (✉)

Department of Psychiatry, School of Medicine, University of California San Diego, 9500 Gilman Drive, M/C 0603, La Jolla, CA 92093-0603, USA

e-mail: amarkou@ucsd.edu

P. Skolnick (ed.), *Glutamate-based Therapies for Psychiatric Disorders*,
Milestones in Drug Therapy, DOI 10.1007/978-3-0346-0241-9_8,
© Springer Basel AG 2010

acute withdrawal syndrome has dissipated [1]. Experimental findings suggest that the neuronal mechanisms underlying the reinforcing properties of drugs of abuse are different from those mediating relapse vulnerability during abstinence [2–4]. Therefore, exploring the effects of putative therapeutic compounds on the different aspects of drug dependence using appropriate animal models is important. Medications targeting the reinforcing effects of drugs of abuse, alleviating the acute withdrawal syndrome, or preventing relapse to drug seeking and drug taking during protracted abstinence could be useful in the treatment of drug and alcohol dependence. This chapter focuses on metabotropic glutamate receptors (mGluRs) as potential treatment targets for drug and alcohol dependence. The role of mGluR in drug dependence has been investigated extensively over the past few years in experimental animal models of drug dependence [5–7], and these findings are summarized below.

2 Neurosubstrates Involved in Drug Dependence and Relapse

Several neurobiological circuits have been shown to be involved in the changes associated with the development of drug dependence. The mesocorticolimbic dopamine system is a crucial mediator of the positive reinforcing effects of drugs of abuse [2, 8–11], although it does not act in isolation. A component of this system is the dopaminergic projection from the ventral tegmental area (VTA) to the nucleus accumbens, amygdala, and frontal cortex [8, 10, 11]. The activity of VTA dopamine neurons is regulated by the release of the excitatory neurotransmitter glutamate which is released from projections originating from several sites, including the nucleus accumbens and frontal cortex [2, 12]. Disruption of the neural system involved in positive reinforcement may also be involved in the negative affective state of early withdrawal [13–15]. With the development of drug dependence, the ability to inhibit drug-seeking behavior in response to environmental cues is impaired. Activation of the prefrontal cortex and amygdala is observed in brain imaging studies in human addicts after exposure to drug-associated cues, leading to increased drug craving [16]. In animal studies, glutamatergic projections from the prefrontal cortex to the extended amygdala and nucleus accumbens have been shown to be involved in drug-induced reinstatement of drug-seeking behavior, whereas projections from the basolateral amygdala to the extended amygdala and nucleus accumbens have been shown to be involved in cue-induced drug-seeking behavior [9, 11, 17]. Other inputs, including γ-aminobutyric acid (GABA) inhibitory interneurons located within the VTA and nucleus accumbens and cholinergic projections from brainstem nuclei to the VTA, are also involved in drug dependence processes [18–20] but will not be discussed in this chapter.

3　Glutamatergic Transmission in Drug Dependence

Glutamatergic neurotransmission plays a critical role in the development of drug dependence [2, 12]. The reinforcing effects of drugs of abuse are likely to involve dopamine and glutamate release in the corticostriatal circuit [6]. The transition from recreational drug use to addiction is associated with the development of compulsive drug use. A decrease in basal glutamate levels was found after chronic exposure to drug self-administration, accompanied by enhanced release of glutamate during reinstatement of drug seeking in rats [6, 17]. These alterations in glutamate transmission may be involved in compulsive drug seeking in humans. Therefore, medications that decrease glutamatergic transmission may reduce the reinforcing and motivational effects of drugs of abuse and alcohol and prevent drug-seeking behavior leading to relapse in humans. By contrast, medications that increase glutamate release may ameliorate early drug withdrawal symptoms associated with decreased glutamatergic transmission after the cessation of chronic drug exposure. Recently, the glutamate homeostasis hypothesis of addiction was proposed [21], in which the maintenance of glutamate homeostasis is based on the balance between glial and synaptic glutamate release and elimination [21]. This balance affects synaptic activity and plasticity by controlling glutamate access to ionotropic and mGluRs.

4　Metabotropic Glutamate Receptors and Glutamatergic Neurotransmission

mGluRs are classified into Groups I–III, based on sequence homology, transduction mechanisms, and pharmacology [22]. Group I mGluRs (mGluR1 and mGluR5) are coupled via G_q proteins to phospholipase C and are located primarily postsynaptically, where they positively mediate the excitatory effects of glutamate on N-methyl-D-aspartate (NMDA) receptors [23–25]. Administration of an mGluR1 agonist increased glutamate transmission in the nucleus accumbens [26], whereas blockade of mGluR5 decreased glutamate neurotransmission [27]. Thus, mGluR5 antagonists may be useful treatments in aspects of drug and alcohol dependence associated with increased glutamate release (e.g., reinforcing and motivational effects of drugs of abuse and drug-seeking behavior). Furthermore, Group II (mGluR2 and mGluR3) and Group III (mGluR4–mGluR8) receptors are coupled to G_i proteins and inhibit adenylate cyclase activity. The release of both glutamate and GABA is regulated by Group III mGluRs [28]. mGluR2 and mGluR3 function as inhibitory autoreceptors that regulate glutamate release or presynaptic heteroceptors that control the release of neurotransmitters other than glutamate [23]. Activation of mGluR2 and mGluR3 reduces evoked glutamate release and decreases glutamate-mediated excitation of postsynaptic receptors [23]. Thus, similar to mGluR5 antagonists, mGluR2/3 agonists may decrease the reinforcing effects of drugs of abuse and decrease drug-seeking behavior by reducing glutamatergic transmission. mGluR2 and mGluR3 are highly expressed in forebrain regions and are mainly located presynaptically outside of the active axon terminal [23]. This extrasynaptic location suggests that mGluR2/3 regulates and prevents high glutamate excitation, which could be pathological, but may not interfere with physiological transmission [23, 29]. One of the approaches to the pharmacologic regulation of glutamate release involves the stimulation of mGlu2/3 autoreceptors by administration of mGluR2/3 agonists or indirectly by increasing extracellular glutamate levels using N-acetylcysteine, which is an amino acid and cysteine prodrug. In the brain, extracellular cysteine is exchanged for intracellular glutamate to maintain extracellular glutamate levels [21, 30]. Moreover, neurochemical analyses revealed that mGluR2/3 function in corticolimbic rat brain sites, including the VTA and nucleus accumbens, was decreased during early drug withdrawal (e.g., nicotine withdrawal), demonstrated by decreased coupling of mGlu2/3 receptors to G proteins in the GTPγ^{35}S binding assay [31], and increased expression of NMDA receptor subunits in subcortical brain sites was observed [32]. These findings reflect a putative compensatory response to the hypothesized decreased glutamate levels during early psychostimulant withdrawal, such as nicotine withdrawal (for review, see [5, 33]).

In summary, as shown by the data below, postsynaptic mGluR5 and presynaptic mGluR2/3 may be particularly interesting as potential targets for medications to treat drug dependence in general and psychostimulant dependence in particular. A number of mGluR agonists and antagonists have been synthesized and tested in animal models of drug dependence (Table 1).

Table 1 List of mGluR compounds used in animal models of drug dependence

Receptor	Action	Abbreviation	Full chemical name
mGluR5	Antagonist	MPEP	2-methyl-6-(phenylethynyl)-pyridine
mGluR5	Antagonist	MTEP	([2-methyl-1,3-thiazol-4-yl]ethynyl)pyridine
mGluR1	Antagonist	AIDA	(R,S)-1-aminoindan-1,5-dicarboxylic acid
mGluR1	Antagonist	EMQMCM	JNJ16567083 (3-ethyl-2-methyl-quinolin-6-yl-[4-methoxy-cyclohexyl]-methanone methanesulfonate)
mGluR1	Antagonist	CPCCOEt	7-(hydroxyimino)cyclopropa[b]chromen-1a-carboxylate ethyl ester.
mGluR2/3	Antagonist	LY341495	$(2S)$-2-amino-2-([1S,2S]-2-carboxycycloprop-1-yl)-3-(xanth-9-yl) propionic acid
mGluR2	Antagonist	MCC	2S,1′S,2′S-2-methyl-2-(2′-carboxycyclopropyl) glycine G
mGluR3	Antagonist	MAP4	α-methyl-L-amino-4-phosphonobutanoate
mGluR2/3	Agonist	LY379268	$(-)$-2-oxa-4-aminobicyclo(3.1.0)hexane-4,6-dicarboxylate
mGluR2/3	Agonist	LY314582	A racemic mixture of LY354740 ([+]-2-aminobicyclo[3.1.0]hexane-2,6-dicarboxylic acid)
mGluR2/3	Agonist	DCG-IV	$(2S,1′R,2′R,3′R)$-2-(2′,3′-dicarboxycyclopropyl) glycine
mGluR2/3	Agonist	(S)-3,4-DCPG	(S)-3,4-dicarboxyphenylglycine

5 Role of mGluRs in Preclinical Models of Drug Dependence

5.1 Role of mGluRs in the Modulation of the Reinforcing and Motivational Effects of Drugs of Abuse

A source of motivation that contributes to the maintenance of drug abuse is the rewarding effect of drugs of abuse. The intravenous drug self-administration procedure in rodents and primates provides a reliable and robust model of human drug consumption studied in controlled laboratory settings [34, 35]. Self-administration under fixed-ratio schedules of reinforcement provides an operational measure of the primary rewarding effects of drugs of abuse and drug intake. Self-administration under the progressive-ratio schedule of reinforcement primarily assesses the motivation for the drug (e.g., [36, 37]).

A plethora of recent findings indicated that mGluRs, particularly Group I and Group II, are involved in mediating drug reward and reinforcement. After an initial study by Chiamulera and colleagues showing that mice lacking mGluR5 do not self-administer cocaine [38], numerous studies demonstrated that mGluR5 antagonists, such as MPEP or MTEP, reduced self-administration of cocaine [39–43] and nicotine [39, 41, 44–47] under fixed-ratio schedules of reinforcement (Fig. 1a, b). Furthermore, in a variety of mouse and rat strains, MPEP and MTEP decreased the reinforcing effects of self-administered alcohol measured using both operant [48–53] (Fig. 1c) and bottle-choice [49, 50, 54] procedures. Additionally, MPEP inhibited the discriminative stimulus properties of consumed alcohol during a self-administration test session [55]. Similar to mGluR5 antagonists, the mGluR1 antagonist CPCCOEt reduced operant alcohol self-administration ([49]; but see [48]). In contrast to psychostimulants and alcohol, MPEP had a small effect (20%) on heroin reinforcement, reflected in the self-administration procedure [56]. Interestingly, MPEP or MTEP had no effect on responding for food pellets [44, 46, 56] (Fig. 1b) or condensed milk [43] under fixed-ratio schedules of reinforcement.

MPEP also decreased the motivation to take a drug, demonstrated by decreased progressive-ratio breakpoints for cocaine, nicotine [45], and alcohol [51]. However, this effect was not drug-specific. MPEP also decreased responding for food under the progressive-ratio schedule [45].

The mGluR2/3 agonist LY379268 decreased cocaine self-administration in rats [57] (Fig. 1d) and squirrel monkeys [58]. LY379268 administered systemically or centrally into the nucleus accumbens or VTA decreased self-administration of nicotine [31] and alcohol [59] (Fig. 1e, f). Administration of this same mGluR2/3 agonist attenuated enhanced amphetamine self-administration in amphetamine-sensitized rats [60]. Interestingly, LY379268 had no effect on heroin self-administration [61], similar to the small effects of an mGluR5 antagonist on heroin self-administration [56]. These findings indicate that activation of mGluR2/3 may be effective specifically against the reinforcing properties of psychostimulant drugs, but not opiates.

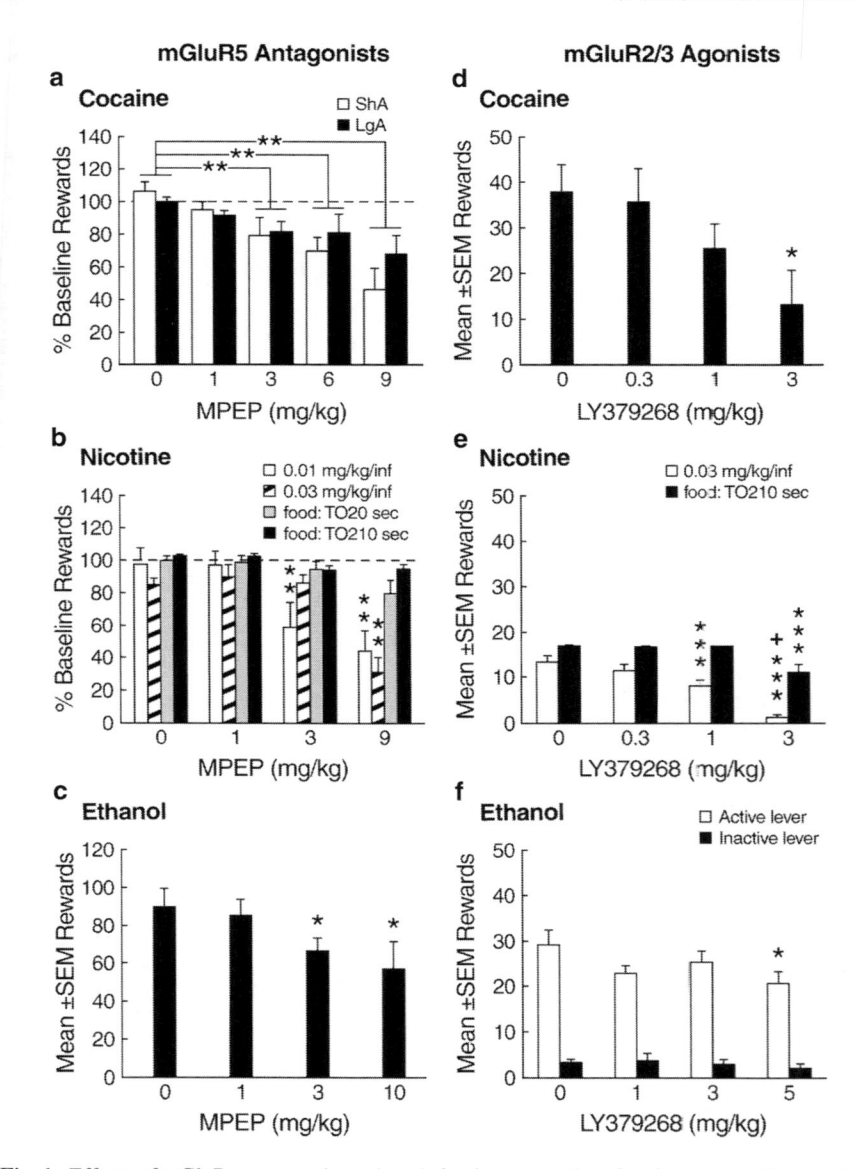

Fig. 1 Effects of mGluR compounds on the reinforcing properties of various drugs of abuse. Both mGluR5 antagonists (MPEP and MTEP, left panels) and an mGluR2/3 agonist (LY379268, right panels) decreased self-administration of cocaine (**a** and **d**), nicotine (**b** and **e**), and alcohol (**c** and **f**) in rats. Figures are taken with permission from [40], [44], [49], [57], [31], and [59], respectively. Please note that direct comparisons of magnitudes of the effects cannot be made among figures because different self-administration parameters were used to conduct the studies reported here

Although activation of mGluR2/3 was effective in attenuating the reinforcing effects of drugs of abuse, these compounds also decreased responding for food pellets (e.g., Fig. 1e), suggesting decreased motivation for a natural reinforcer. This effect was more pronounced when food responding was tested in food-restricted animals or when less palatable food was used as a reinforcer [45, 46]. However, other literature findings reported that LY379268 (1 or 3 mg/kg) had no effect on sucrose intake in free-feeding rats [57] or responding for condensed milk in food-restricted rats [62]. These findings suggest that higher doses of LY379268 may be needed to decrease responding for food when animals have low motivational drive, which is the case in rats that are freely fed or work for a palatable reinforcer under no deprivation conditions when they are not food-deprived.

The effects of blockade of mGluR2/3 in drug and alcohol dependence are less clear. Administration of the mGluR2/3 antagonist LY341495 decreased nicotine self-administration and food-maintained responding [46], suggesting that this effect may be nonspecific. Furthermore, LY341495 had no effect on alcohol self-administration even when used at a high dose of 10 mg/kg [52]. Considering the consistent effects of mGluR2/3 agonists in decreasing the reinforcing and motivational effects of drugs of abuse (see above), it is not surprising that antagonists at the same receptors appear to have no clear effects on drug self-administration.

Administration of (S)-3,4-DCPG, an agonist at presynaptic mGluR8 that negatively modulates glutamate transmission [63], decreased alcohol self-administration at doses that also decreased spontaneous locomotor activity [59]. Thus, this effect of an mGluR8 agonist on the reinforcing effect of alcohol may be nonspecific and attributed to general behavioral suppression.

Altogether, these studies in experimental animals showed that blockade of postsynaptic mGluR5 and activation of presynaptic mGluR2/3 resulted in attenuation of both the reinforcing and motivational effects of drugs of abuse, including psychostimulants (nicotine and cocaine) and alcohol (Fig. 1), with no effect on opiate reinforcement. Thus, mGluR5 antagonists and mGluR2/3 agonists are expected to be effective treatments during the phase of drug dependence associated with increased glutamate transmission. This efficacy may be attributable to the decrease in the primary rewarding effects of drugs of abuse in humans by blocking the increases in glutamate transmission induced by self-administration of psychostimulants or alcohol, but not opiates. Importantly, the effects of mGluR5 blockade were drug-specific and did not affect responding for natural reinforcers, with the exception of cases in which a progressive-ratio schedule of reinforcement was used, which assessed incentive motivation for reinforcers [45]. By contrast, mGluR2/3 agonists decreased the reinforcing and motivational properties of both drugs and food, a natural reinforcer. This potential "side effect" of mGluR2/3 agonists may be diminished by using positive allosteric modulators of mGluR2/3. Allosteric potentiators of mGluR2 have been developed that selectively potentiate the effects of glutamate and other receptor agonists to activate mGluR2 [64–66]. Preliminary findings with the mGluR2 positive allosteric modulator BINA [67, 68] showed that it attenuated the reinforcing effects of cocaine and decreased cue-induced cocaine-seeking behavior without affecting behaviors motivated by food reinforcement (Semenova and Markou, unpublished observations).

5.2 Role of mGluRs in the Modulation of the Reward-Enhancing Effects of Drugs of Abuse

Another source of motivation for drug use and abuse is the reward-enhancing effects of drugs of abuse. Drugs of abuse amplify reward signals in the brain elicited by other reinforcers, including natural reinforcers [69–72]. This action of drugs of abuse may partially account for their intrinsic rewarding properties and also may explain how psychostimulants increase sensitivity to nondrug rewarding environmental stimuli [73–76]. One method for assessing the reward-enhancing properties of drugs of abuse involves the intracranial self-stimulation (ICSS) procedure. Brief electrical pulses to discrete regions of the brain reward circuit are extremely reinforcing for rats, which will perform an operant task to self-deliver the stimulation [77]. By systematically varying a parameter of the stimulation, such as current–intensity, one may derive thresholds that reflect the functioning of the brain reward system. Administration of various drugs of abuse lowers ICSS thresholds in rats [78], reflecting drug-induced enhancement of the rewarding effects of the stimulation. Investigations of the effects of mGluR compounds on the reward-enhancing effects of drugs of abuse are currently limited to studies with nicotine. Specifically, both an mGluR5 antagonist and an mGluR2/3 agonist blocked the reward-enhancing effects of nicotine [76, 79], but this "blockade" of the reward-enhancing effects of nicotine may be additive and therefore nonspecific. That is, nicotine enhanced brain reward function, and the glutamate compounds decreased brain reward function, resulting in an apparent blockade of the reward-enhancing effects of nicotine. Although these effects appear to be pharmacologically additive, when coupled with the ability of these compounds to decrease nicotine self-administration, this blockade of the effects of nicotine on reward enhancement may be clinically relevant. These compounds may block the reward-enhancing effects of drugs of abuse and thus remove a source of motivation to use drugs, such as nicotine. Further studies are needed to evaluate the effects of mGluR compounds on the reward-enhancing effects of other drugs of abuse, including alcohol.

5.3 Role of mGluRs in the Modulation of the Conditioned Rewarding Effects of Drugs of Abuse

An animal model of the conditioned rewarding effects of drugs of abuse that is widely used is the conditioned place preference (CPP) procedure [80]. This test utilizes Pavlovian conditioning procedures to measure the "subjective" rewarding effects of drugs of abuse based on the animal's preference for an environment associated with drug administration over a nondrug-associated environment. CPP does not directly measure primary drug reinforcement, but rather the motivation for a secondary reinforcer expressed as preference for the drug-paired environment. Blockade of the rewarding effects of a drug can be assumed if a drug blocks the acquisition or expression of CPP. MPEP decreased the conditioned rewarding effects of cocaine [81] and methamphetamine [82] in mice during the acquisition of CPP. MPEP also blocked the expression of amphetamine-induced CPP in rats [83]. Although MPEP had no effects on the acquisition of CPP to D-amphetamine or nicotine [81], the lack of effects of MPEP on CPP could be attributed to the low MPEP dose used in these studies. MPEP attenuated the expression [49], but not acquisition [81], of alcohol-induced CPP, suggesting a possible nonspecific effect, because other studies showed that MPEP decreased alcohol self-administration [48–54].

The effects of MPEP on opiate-induced CPP were shown to be bidirectional. Specifically, moderate MPEP doses (10 mg/kg) potentiated heroin-induced CPP in rats [84], whereas higher MPEP doses (30 and 50 mg/kg) or intracerebroventricular MPEP administration decreased the conditioned rewarding effects of morphine in mice [85, 86] and rats [87]. Nevertheless, attenuation of CPP by the highest MPEP doses may be nonspecific and attributable to behavioral suppression [88].

The effects of mGluR2/3 agonists or antagonists on the conditioned rewarding effects of drugs of abuse remain to be evaluated. Other pharmacological approaches that modulate glutamatergic transmission were effective in reducing the conditioned rewarding effects of drugs of abuse. MS-153, a glutamate transporter activator, attenuated the acquisition of CPP to morphine, methamphetamine, and cocaine in mice [89]. Furthermore, N-acetylated-α-linked-acidic dipeptidase inhibitors prevented the acquisition and expression of morphine- and cocaine-induced CPP [90, 91].

In summary, these findings indicate that, similar to the attenuation of the reinforcing effects of drugs of abuse, blockade of mGluR5 attenuated the conditioned rewarding effects of psychostimulant drugs and alcohol, but not opiates, possibly by blocking increased glutamate transmission at mGluR5 induced by administration of the drug of abuse.

5.4 Role of mGluRs in the Modulation of Different Aspects of Drug Withdrawal, Including Anhedonia, Depression, and Anxiety

Drug withdrawal is associated with negative affective symptoms in humans, including anhedonia, depressed mood, irritability, craving, and anxiety [1]. A somatic withdrawal syndrome has also been characterized in both humans and rodents [1, 92, 93]. This aversive abstinence syndrome is hypothesized to contribute to the persistence of drug taking and relapse during abstinence [9, 11, 94–97]. Therefore, amelioration of drug withdrawal with medications may prevent relapse to drug taking during the early withdrawal phase when the urge to reinitiate drug use is the strongest [96, 98–102].

In laboratory animals, withdrawal from administration of different drugs of abuse results in a deficit in brain reward function, measured by elevations in ICSS reward thresholds [77, 78]. Such elevations of brain reward thresholds are an operational measure of "diminished interest or pleasure" in rewarding stimuli (i.e., anhedonia). Acute early withdrawal from chronic cocaine [103], amphetamine [104, 105], opiates [106], nicotine [93], and ethanol [107] led to elevated brain reward thresholds. Interestingly, a clear dissociation can be found between the depression-like anhedonic aspects of nicotine withdrawal and the somatic aspects of nicotine withdrawal, suggesting that different mechanisms underlie these phenomena [108, 109]. The anhedonic aspects of nicotine withdrawal have been suggested to be centrally mediated, whereas the somatic aspects of nicotine withdrawal are peripherally mediated [108]. Therefore, assessing the effects of treatments on both the affective depression-like and somatic aspects of drug withdrawal is important, although the affective aspects of drug withdrawal are hypothesized to be more critically involved in drug dependence than the somatic signs of withdrawal [9, 14].

The mGluR5 antagonist MPEP significantly elevated brain reward thresholds and increased somatic withdrawal signs associated with spontaneous nicotine withdrawal [46], but it did so equally in control rats [39, 76], indicating that MPEP nonspecifically decreased brain reward function and increased the number of somatic signs. Thus, mGluR5 antagonism may be expected to worsen symptoms of early nicotine withdrawal and withdrawal from other psychomotor stimulant drugs that are hypothesized to be accompanied by decreased glutamate transmission [5]. By contrast, during opiate withdrawal, which is associated with increased glutamate transmission [110], MTEP and MPEP attenuated naloxone-induced somatic signs of morphine withdrawal in mice [111] and rats [112]. The effects of mGluR5 antagonists on the affective aspects of opiate withdrawal have not been studied.

The severity of the somatic signs observed in rats chronically exposed to morphine and injected acutely with an opiate receptor antagonist was attenuated by administration of the mGluR2/3 agonists DCG-IV [113] and LY354740 [114, 115]. The effects of mGluR2/3 agonists on the affective aspects of opiate withdrawal are not known. Moreover, administration of the mGluR2/3 agonist LY379268 tended to aggravate nicotine withdrawal-induced reward deficits, whereas it produced mild reward deficits in control rats [116]. Thus, mGlu2/3 agonists did not appear to significantly influence the affective anhedonic-like aspects of nicotine withdrawal but attenuated the somatic aspects of morphine withdrawal.

Again, differential effects on nicotine and morphine withdrawal were observed after administration of mGluR Group II antagonists. Early work demonstrated that acute systemic administration of the nonselective mGluR2/3 antagonists MCCG and MAP4, which have actions at other mGluRs, did not alter the somatic signs seen during early opiate withdrawal [117]. Administration of the selective mGluR2/3 antagonist LY341495 reversed affective aspects of nicotine withdrawal [79] but exacerbated somatic signs after administration of an opiate receptor antagonist in morphine-treated rats [118].

Interestingly, attenuation of the depression-like reward deficits of nicotine withdrawal with the mGluR2/3 antagonist LY341495 is consistent with the antidepressant-like effects of LY341495 and MGS0039 in the forced swim test and the learned helpless test in rats and the tail suspension test in mice [119, 120]. Additionally, these findings are consistent with previous reports demonstrating that reward deficits associated with psychostimulant withdrawal were reversed after administration of various antidepressant treatments, indicating that the elevations in ICSS thresholds during psychostimulant withdrawal have predictive validity as an animal model of anhedonia (for reviews, see [121–124]). The antidepressant properties of mGluR2/3 antagonists are discussed elsewhere in this volume.

Altogether, the differential effects of mGluR5 and mGluR2/3 antagonists on nicotine withdrawal are consistent with the hypothesis that nicotine withdrawal is associated with decreased glutamatergic transmission [19, 33]. Accordingly, mGluR5 antagonists resulted in worsening of both the affective and somatic aspects of nicotine withdrawal, whereas an mGluR2/3 antagonist reversed reward deficits associated with nicotine withdrawal. However, unknown is whether these compounds would have similar effects on the affective aspects of psychostimulant withdrawal induced by drugs of abuse other than nicotine. Consistent with the reported increase in glutamate transmission during morphine withdrawal [110], both an mGluR5 antagonist and an mGluR2/3 agonist, but not an mGluR2/3 antagonist, attenuated naloxone-induced somatic signs of opiate withdrawal in opiate-dependent rats. Data on the effects of these compounds on the affective aspects of opiate withdrawal are currently lacking. Thus, based on the hypothesis outlined above, in which the affective and somatic aspects of drug withdrawal may be mediated by differential neurobiological mechanisms, definite conclusions about the differential effects of mGluR5 antagonists and mGluR2/3 agonists and antagonists on psychostimulant and opiate withdrawal cannot yet be made. Notably, nicotine withdrawal was induced by termination of chronic nicotine administration, whereas opiate withdrawal was induced by administration of an opioid receptor antagonist to opiate-dependent subjects. Microdialysis studies are needed to determine the directional changes in glutamate transmission during early psychostimulant and opiate withdrawal. These studies will be greatly important, especially considering recent findings showing differential effects of the same drug treatment administered during spontaneous and antagonist-precipitated psychostimulant withdrawal [125].

Another affective sign of drug withdrawal that is seen after termination of administration of a variety of drugs of abuse is anxiety-like behavior. Different animal models revealed anxiety-like responses during acute withdrawal from cocaine [126, 127], opiates [128, 129], ethanol [130–132], and nicotine [133–135].

Anxiety tests, such as fear- or light-potentiated startle, which assess passive reactivity to stressors, show predictive validity for standard anxiolytic compounds [136]. Recently, fear-potentiated startle was shown to be sensitive to the anxiolytic-like effects of the mGluR2/3 agonist LY354740 and its prodrug LY544344. Specifically, anxiolytic effects of LY354740 were seen in the fear-potentiated startle paradigm in humans [137]. Additionally, LY354740 and LY544344 reduced generalized anxiety disorder symptoms in humans [138]. Importantly, recent data showed that nicotine withdrawal increases startle reactivity only during an anxiogenic-like situation in rats. Nicotine-abstinent humans exhibited increased fear-potentiated startle compared with nonsmokers [139]. In rats, nicotine withdrawal enhanced light-potentiated startle, without altering startle reactivity during the nonstressful dark testing conditions [134]. These findings support the hypothesis that nicotine withdrawal exacerbates stress responding and indicate that fear- or light-potentiated startle tests could provide important information about how mGluR compounds may alter responses to stress during drug withdrawal.

5.5   Role of mGluRs in the Modulation of Drug-Seeking Behavior

Chronic vulnerability to relapse during protracted abstinence is one of the main challenges for the treatment of drug addiction [96, 140]. One of the conditions that precipitates drug craving and relapse to drug use in humans is the presentation of environmental stimuli previously paired with drug taking [141–143]. In animals, stimuli previously associated with drug administration may elicit drug-seeking behaviors, leading to reinstatement of drug self-administration after a prolonged period of abstinence or after self-administration behavior has been extinguished (e.g., [35, 36, 96, 97]). Additionally, drug-seeking behavior may be induced by the administration of the drug of abuse, by other drugs that share discriminatory properties with the original drug of abuse, or by exposure to stress or an environment previously associated with drug delivery.

The mGluR5 antagonists MPEP and MTEP decreased nicotine-induced nicotine-seeking behavior [41] and cue-induced reinstatement of drug-seeking behavior for cocaine [144], nicotine [145], and alcohol [146] (Fig. 2a–c, respectively). Additionally, the mGluR1 antagonist decreased both cue- and nicotine-induced reinstatement of nicotine self-administration [147].

Administration of the mGluR2/3 agonist LY379268 attenuated cue- [57, 148] and cocaine- [58, 149] induced reinstatement of cocaine self-administration in rats (Fig. 2d). Furthermore, systemic or central LY379268 administration attenuated cue-, stress-, or context-induced drug-seeking behavior for heroin [61, 150, 151], nicotine [31], and alcohol [59, 152] in rats (Fig. 2e, f). LY379268 also reduced cue-induced food pellet- [149], condensed milk- [57], and sucrose- [62] seeking behavior, without impairing locomotion [62, 153, 154]. These findings demonstrate that

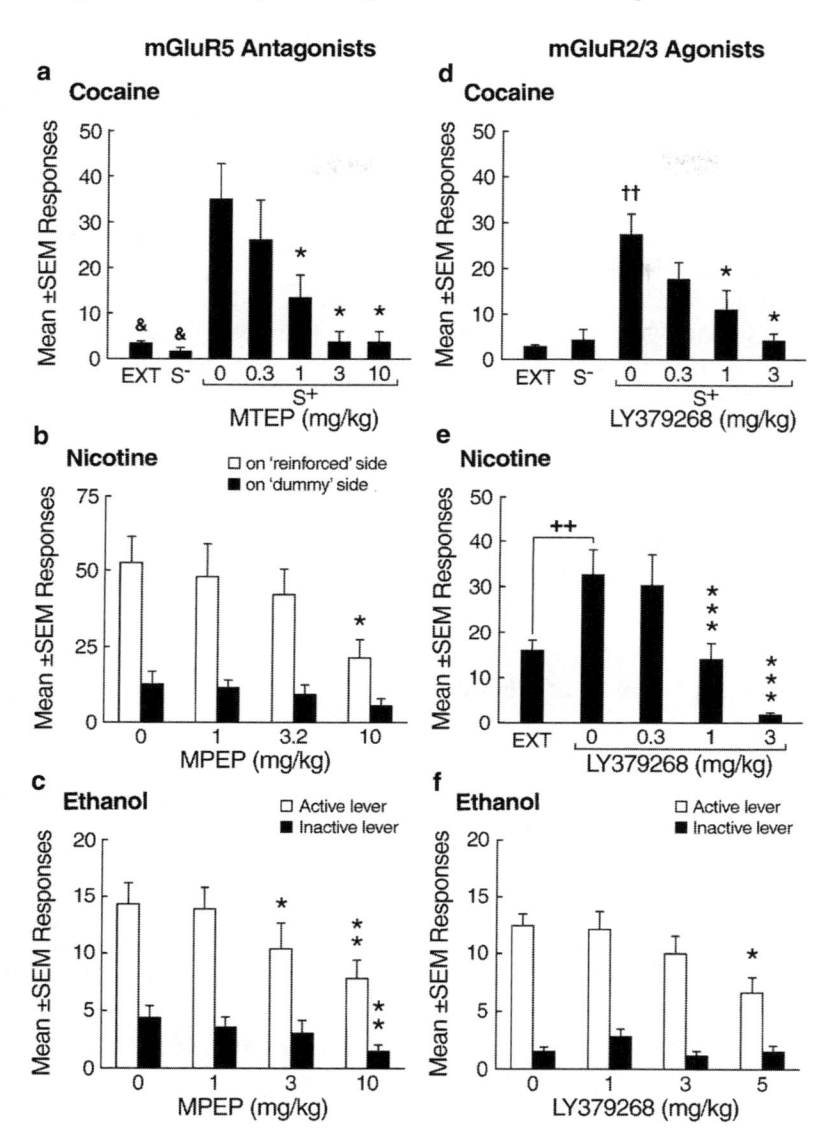

Fig. 2 Effects of mGluR compounds on drug-seeking behavior. Both mGluR5 antagonists (MPEP and MTEP, *left* panels) and an mGluR2/3 agonist (LY379268, *right* panels) abolished reinstatement of extinguished self-administration of cocaine (**a** and **d**), nicotine (**b** and **e**), and alcohol (**c** and **f**) in rats. Figures are taken with permission from [43], [145], [146], [57], [31], and [59], respectively. Please note that direct comparisons of magnitudes of the effects cannot be made among figures because different parameters for drug self-administration, extinction, and reinstatement were used to conduct the studies that generated the data depicted here

administration of an mGluR2/3 agonist decreased the motivation to seek both drug and a natural reinforcer.

The mGluR8 agonist (S)-3,4-DCPG decreased reinstatement of alcohol self-administration at doses that decreased spontaneous locomotor activity [59]. Thus, this effect of the mGluR8 agonist on alcohol-seeking behavior may be nonspecific and attributable to general behavioral suppression.

In rats previously treated with cocaine or self-administering cocaine, administration of N-acetylcysteine restored extracellular glutamate levels in the nucleus accumbens and prevented cocaine-primed reinstatement of cocaine-seeking [155–157]. N-acetylcysteine similarly decreased cue- and heroin-induced heroin seeking [158]. In humans, N-acetylcysteine decreased cue-induced craving for cocaine [159–161] and pathological gambling in a pilot study [162]. Human smokers treated with N-acetylcysteine reported a reduction in cigarettes smoked when alcohol consumption was taken into account, but no effect of N-acetylcysteine was observed on estimates of carbon monoxide levels, craving, or withdrawal [163]. These findings indicate that this pharmacological manipulation of the glutamate system may also help to decrease addictive and compulsive reward-seeking behaviors. These preliminary clinical data in smokers will need further confirmation in larger placebo-controlled trials.

6 Summary and Conclusions

The results from the experimental animal studies summarized in this chapter demonstrated a critical role for glutamate transmission in drug and alcohol dependence and the involvement of various mGluRs in behaviors relevant to several aspects of drug dependence. Based on these preclinical findings, potential candidate medications include postsynaptic mGluR5 antagonists, which inhibit glutamate transmission, and presynaptic mGluR2/3 autoreceptor agonists and antagonists, which inhibit excessive glutamate release or allow glutamate release, respectively. Increases or decreases in glutamate transmission as therapeutic approaches to drug addiction are recommended for the treatment of different aspects of drug addiction. Both mGluR5 antagonists and mGluR2/3 agonists may decrease the rewarding and motivational effects of various psychostimulants (e.g., cocaine, nicotine, amphetamine) and alcohol, but not opiates, and prevent cue-, context-, and drug-induced reinstatement of extinguished self-administration of psychostimulants, opiates, and alcohol by suppressing the increases in glutamate transmission induced by drugs or the presentation of stimuli previously associated with drug effects or the availability of drugs. Nevertheless, in contrast to mGluR5 antagonists, which had mostly no effects on the reinforcing effects of nondrug reinforcers, mGluR2/3 agonists decreased the reinforcing and motivational properties of natural reinforcers, suggesting a potential "side effect" of direct stimulation of mGluR2/3. Preliminary findings with an mGluR2 positive modulator suggest that this may be an improved pharmacological approach for decreasing the reinforcing and motivational effects of drugs of abuse, with minimal effects on natural reinforcers. Data on the effects of mGluR compounds on the affective and somatic aspects of drug withdrawal are less clear. Based on studies with nicotine, mGluR2/3 antagonists may be useful in the treatment of the depression-like affective symptoms of psychostimulant withdrawal by reversing the hypothesized decreases in glutamate transmission during the early nicotine withdrawal phase. By contrast, mGluR5 antagonists and mGluR2/3 agonists showed some efficacy in reversing the naloxone-induced somatic aspects of opiate withdrawal in opiate-dependent subjects that may be associated with increased glutamate transmission. Affective and somatic aspects of drug withdrawal may be mediated by different neurobiological mechanisms. Accordingly, differential effects of mGluR compounds on the affective and somatic aspects of nicotine and morphine withdrawal, respectively, have been observed. Finally, based on the anxiolytic-like effects of mGluR5 and mGluR2/3 compounds in several animal tests of anxiety [164], mGluR5 and mGluR2/3 compounds may be predicted to alleviate the increased anxiety and irritability symptoms associated with the early withdrawal phase in humans [1]. Such additional beneficial effects would further contribute to the treatment of drug dependence.

In conclusion, ample preclinical data strongly suggest the potential clinical utility of mGluR compounds for the treatment of various aspects of drug and alcohol dependence. Clinical testing of mGluR compounds as medications for drug and alcohol dependence is warranted.

Acknowledgments This work was supported by National Institute on Drug Abuse grants (R01) DA11946 and (R01) DA023209 to Athina Markou. Both authors have a patent application regarding the use of mGluR compounds for the treatment of drug addiction. The authors have no other conflicts of interest to declare that are directly relevant to the content of this chapter. The authors would like to thank Mr. Michel Arends for editorial assistance and Ms. Janet Hightower for computer graphics.

References

1. American Psychiatric Association (1994) Diagnostic and statistical manual of mental disorders, 4th edn. American Psychiatric Press, Washington DC
2. Kalivas PW, Volkow ND (2005) The neural basis of addiction: a pathology of motivation and choice. Am J Psychiatry 162:1403–1413
3. Shalev U, Grimm JW, Shaham Y (2002) Neurobiology of relapse to heroin and cocaine seeking: a review. Pharmacol Rev 54:1–42
4. Shiffman S, Ferguson SG, Gwaltney CJ, Balabanis MH, Shadel WG (2006) Reduction of abstinence-induced withdrawal and craving using high-dose nicotine replacement therapy. Psychopharmacology (Berl) 184:637–644
5. Markou A (2007) Metabotropic glutamate receptor antagonists: novel therapeutics for nicotine dependence and depression? Biol Psychiatry 61:17–22
6. Gass JT, Olive MF (2008) Glutamatergic substrates of drug addiction and alcoholism. Biochem Pharmacol 75:218–265
7. Kalivas PW, Lalumiere RT, Knackstedt L, Shen H (2009) Glutamate transmission in addiction. Neuropharmacology 56(Suppl 1):169–173

8. Koob GF, Volkow ND (2010) Neurocircuitry of addiction. Neuropsychopharmacology 35:217–238
9. Koob GF (2009) Neurobiological substrates for the dark side of compulsivity in addiction. Neuropharmacology 56(Suppl 1):18–31
10. Nestler EJ (2005) Is there a common molecular pathway for addiction? Nat Neurosci 8:1445–1449
11. Feltenstein MW, See RE (2008) The neurocircuitry of addiction: an overview. Br J Pharmacol 154:261–274
12. Kalivas PW, Volkow N, Seamans J (2005) Unmanageable motivation in addiction: a pathology in prefrontal-accumbens glutamate transmission. Neuron 45:647–650
13. Koob GF (2008) A role for brain stress systems in addiction. Neuron 59:11–34
14. Koob GF, Le Moal M (2008) Neurobiological mechanisms for opponent motivational processes in addiction. Philos Trans R Soc Lond B Biol Sci 363:3113–3123
15. Koob GF (2003) Neuroadaptive mechanisms of addiction: studies on the extended amygdala. Eur Neuropsychopharmacol 13:442–452
16. Goldstein RZ, Volkow ND (2002) Drug addiction and its underlying neurobiological basis: neuroimaging evidence for the involvement of the frontal cortex. Am J Psychiatry 159:1642–1652
17. Knackstedt LA, Kalivas PW (2009) Glutamate and reinstatement. Curr Opin Pharmacol 9:59–64
18. Mansvelder HD, Keath JR, McGehee DS (2002) Synaptic mechanisms underlie nicotine-induced excitability of brain reward areas. Neuron 33:905–919
19. Markou A (2008) Neurobiology of nicotine dependence. Philos Trans R Soc Lond B Biol Sci 363:3159–3168
20. Markou A, Paterson NE, Semenova S (2004) Role of γ-aminobutyric acid (GABA) and metabotropic glutamate receptors in nicotine reinforcement: potential pharmacotherapies for smoking cessation. Ann N Y Acad Sci 1025:491–503
21. Kalivas PW (2009) The glutamate homeostasis hypothesis of addiction. Nat Rev Neurosci 10:561–572
22. Pin JP, Duvoisin R (1995) The metabotropic glutamate receptors: structure and functions. Neuropharmacology 34:1–26
23. Schoepp DD (2001) Unveiling the functions of presynaptic metabotropic glutamate receptors in the central nervous system. J Pharmacol Exp Ther 299:12–20
24. Awad H, Hubert GW, Smith Y, Levey AI, Conn PJ (2000) Activation of metabotropic glutamate receptor 5 has direct excitatory effects and potentiates NMDA receptor currents in neurons of the subthalamic nucleus. J Neurosci 20:7871–7879
25. Pisani A, Gubellini P, Bonsi P, Conquet F, Picconi B, Centonze D, Bernardi G, Calabresi P (2001) Metabotropic glutamate receptor 5 mediates the potentiation of N-methyl-D-aspartate responses in medium spiny striatal neurons. Neuroscience 106:579–587
26. Swanson CJ, Baker DA, Carson D, Worley PF, Kalivas PW (2001) Repeated cocaine administration attenuates group I metabotropic glutamate receptor-mediated glutamate release and behavioral activation: a potential role for Homer. J Neurosci 21:9043–9052
27. Attucci S, Carla V, Mannaioni G, Moroni F (2001) Activation of type 5 metabotropic glutamate receptors enhances NMDA responses in mice cortical wedges. Br J Pharmacol 132:799–806
28. Snead OC 3rd, Banerjee PK, Burnham M, Hampson D (2000) Modulation of absence seizures by the GABA_A receptor: a critical role for metabotropic glutamate receptor 4 (mGluR4). J Neurosci 20:6218–6224
29. Cartmell J, Schoepp DD (2000) Regulation of neurotransmitter release by metabotropic glutamate receptors. J Neurochem 75:889–907
30. Melendez RI, Vuthiganon J, Kalivas PW (2005) Regulation of extracellular glutamate in the prefrontal cortex: focus on the cystine glutamate exchanger and group I metabotropic glutamate receptors. J Pharmacol Exp Ther 314:139–147

31. Liechti ME, Lhuillier L, Kaupmann K, Markou A (2007) Metabotropic glutamate 2/3 receptors in the ventral tegmental area and the nucleus accumbens shell are involved in behaviors relating to nicotine dependence. J Neurosci 27:9077–9085

32. Kenny PJ, Chartoff E, Roberto M, Carlezon WA Jr, Markou A (2009) NMDA receptors regulate nicotine-enhanced brain reward function and intravenous nicotine self-administration: role of the ventral tegmental area and central nucleus of the amygdala. Neuropsychopharmacology 34:266–281

33. Liechti ME, Markou A (2008) Role of the glutamatergic system in nicotine dependence: implications for the discovery and development of new pharmacological smoking cessation therapies. CNS Drugs 22:705–724

34. Caine SB, Koob GF (1993) Intravenous drug self-administration techniques in animals. In: Sahgal A (ed) Behavioural neuroscience: a practical approach, vol 2. IRL Press, Oxford, pp 117–143

35. Katz JL, Higgins ST (2003) The validity of the reinstatement model of craving and relapse to drug use. Psychopharmacology (Berl) 168:21–30

36. Markou A, Weiss F, Gold LH, Caine SB, Schulteis G, Koob GF (1993) Animal models of drug craving. Psychopharmacology (Berl) 112:163–182

37. Arnold JM, Roberts DC (1997) A critique of fixed and progressive ratio schedules used to examine the neural substrates of drug reinforcement. Pharmacol Biochem Behav 57:441–447

38. Chiamulera C, Epping-Jordan MP, Zocchi A, Marcon C, Cottiny C, Tacconi S, Corsi M, Orzi F, Conquet F (2001) Reinforcing and locomotor stimulant effects of cocaine are absent in mGluR5 null mutant mice. Nat Neurosci 4:873–874

39. Kenny PJ, Paterson NE, Boutrel B, Semenova S, Harrison AA, Gasparini F, Koob GF, Skoubis PD, Markou A (2003) Metabotropic glutamate 5 receptor antagonist MPEP decreased nicotine and cocaine self-administration but not nicotine and cocaine-induced facilitation of brain reward function in rats. Ann N Y Acad Sci 1003:415–418

40. Kenny PJ, Boutrel B, Gasparini F, Koob GF, Markou A (2005) Metabotropic glutamate 5 receptor blockade may attenuate cocaine self-administration by decreasing brain reward function in rats. Psychopharmacology (Berl) 179:247–254

41. Tessari M, Pilla M, Andreoli M, Hutcheson DM, Heidbreder CA (2004) Antagonism at metabotropic glutamate 5 receptors inhibits nicotine- and cocaine-taking behaviours and prevents nicotine-triggered relapse to nicotine-seeking. Eur J Pharmacol 499:121–133

42. Lee B, Platt DM, Rowlett JK, Adewale AS, Spealman RD (2005) Attenuation of behavioral effects of cocaine by the metabotropic glutamate receptor 5 antagonist 2-methyl-6-(phenylethynyl)-pyridine in squirrel monkeys: comparison with dizocilpine. J Pharmacol Exp Ther 312:1232–1240

43. Martin-Fardon R, Baptista MA, Dayas CV, Weiss F (2009) Dissociation of the effects of MTEP [3-[(2-methyl-1, 3-thiazol-4-yl)ethynyl]piperidine] on conditioned reinstatement and reinforcement: comparison between cocaine and a conventional reinforcer. J Pharmacol Exp Ther 329:1084–1090

44. Paterson NE, Semenova S, Gasparini F, Markou A (2003) The mGluR5 antagonist MPEP decreased nicotine self-administration in rats and mice. Psychopharmacology (Berl) 167:257–264

45. Paterson NE, Markou A (2005) The metabotropic glutamate receptor 5 antagonist MPEP decreased break points for nicotine, cocaine and food in rats. Psychopharmacology (Berl) 179:255–261

46. Liechti ME, Markou A (2007) Interactive effects of the mGlu5 receptor antagonist MPEP and the mGlu2/3 receptor antagonist LY341495 on nicotine self-administration and reward deficits associated with nicotine withdrawal in rats. Eur J Pharmacol 554:164–174

47. Palmatier MI, Liu X, Donny EC, Caggiula AR, Sved AF (2008) Metabotropic glutamate 5 receptor (mGluR5) antagonists decrease nicotine seeking, but do not affect the reinforcement enhancing effects of nicotine. Neuropsychopharmacology 33:2139–2147

48. Hodge CW, Miles MF, Sharko AC, Stevenson RA, Hillmann JR, Lepoutre V, Besheer J, Schroeder JP (2006) The mGluR5 antagonist MPEP selectively inhibits the onset and maintenance of ethanol self-administration in C57BL/6J mice. Psychopharmacology (Berl) 183:429–438

49. Lominac KD, Kapasova Z, Hannun RA, Patterson C, Middaugh LD, Szumlinski KK (2006) Behavioral and neurochemical interactions between Group 1 mGluR antagonists and ethanol: potential insight into their anti-addictive properties. Drug Alcohol Depend 85:142–156

50. Cowen MS, Djouma E, Lawrence AJ (2005) The metabotropic glutamate 5 receptor antagonist 3-[(2-methyl-1, 3-thiazol-4-yl)ethynyl]-pyridine reduces ethanol self-administration in multiple strains of alcohol-preferring rats and regulates olfactory glutamatergic systems. J Pharmacol Exp Ther 315:590–600

51. Cowen MS, Krstew E, Lawrence AJ (2007) Assessing appetitive and consummatory phases of ethanol self-administration in C57BL/6J mice under operant conditions: regulation by mGlu5 receptor antagonism. Psychopharmacology (Berl) 190:21–29

52. Schroeder JP, Overstreet DH, Hodge CW (2005) The mGluR5 antagonist MPEP decreases operant ethanol self-administration during maintenance and after repeated alcohol deprivations in alcohol-preferring (P) rats. Psychopharmacology (Berl) 179:262–270

53. Olive MF, McGeehan AJ, Kinder JR, McMahon T, Hodge CW, Janak PH, Messing RO (2005) The mGluR5 antagonist 6-methyl-2-(phenylethynyl)pyridine decreases ethanol consumption via a protein kinase Cε-dependent mechanism. Mol Pharmacol 67:349–355

54. McMillen BA, Crawford MS, Kulers CM, Williams HL (2005) Effects of a metabotropic, mGlu5, glutamate receptor antagonist on ethanol consumption by genetic drinking rats. Alcohol Alcohol 40:494–497

55. Besheer J, Stevenson RA, Hodge CW (2006) mGlu5 receptors are involved in the discriminative stimulus effects of self-administered ethanol in rats. Eur J Pharmacol 551:71–75

56. van der Kam EL, de Vry J, Tzschentke TM (2007) Effect of 2-methyl-6-(phenylethynyl) pyridine on intravenous self-administration of ketamine and heroin in the rat. Behav Pharmacol 18:717–724

57. Baptista MA, Martin-Fardon R, Weiss F (2004) Preferential effects of the metabotropic glutamate 2/3 receptor agonist LY379268 on conditioned reinstatement versus primary reinforcement: comparison between cocaine and a potent conventional reinforcer. J Neurosci 24:4723–4727

58. Adewale AS, Platt DM, Spealman RD (2006) Pharmacological stimulation of group II metabotropic glutamate receptors reduces cocaine self-administration and cocaine-induced reinstatement of drug seeking in squirrel monkeys. J Pharmacol Exp Ther 318:922–931

59. Backstrom P, Hyytia P (2005) Suppression of alcohol self-administration and cue-induced reinstatement of alcohol seeking by the mGlu2/3 receptor agonist LY379268 and the mGlu8 receptor agonist (S)-3, 4-DCPG. Eur J Pharmacol 528:110–118

60. Kim JH, Austin JD, Tanabe L, Creekmore E, Vezina P (2005) Activation of group II mGlu receptors blocks the enhanced drug taking induced by previous exposure to amphetamine. Eur J Neurosci 21:295–300, erratum: 25:908

61. Bossert JM, Busch RF, Gray SM (2005) The novel mGluR2/3 agonist LY379268 attenuates cue-induced reinstatement of heroin seeking. NeuroReport 16:1013–1016

62. Bossert JM, Poles GC, Sheffler-Collins SI, Ghitza UE (2006) The mGluR2/3 agonist LY379268 attenuates context- and discrete cue-induced reinstatement of sucrose seeking but not sucrose self-administration in rats. Behav Brain Res 173:148–152

63. Thomas NK, Wright RA, Howson PA, Kingston AE, Schoepp DD, Jane DE (2001) (S)-3, 4-DCPG, a potent and selective mGlu8a receptor agonist, activates metabotropic glutamate receptors on primary afferent terminals in the neonatal rat spinal cord. Neuropharmacology 40:311–318

64. Johnson MP, Baez M, Jagdmann GE Jr, Britton TC, Large TH, Callagaro DO, Tizzano JP, Monn JA, Schoepp DD (2003) Discovery of allosteric potentiators for the metabotropic

glutamate 2 receptor: synthesis and subtype selectivity of N-(4-(2-methoxyphenoxy)phe-nyl)-N-(2, 2, 2- trifluoroethylsulfonyl)pyrid-3-ylmethylamine. J Med Chem 46:3189–3192

65. Pinkerton AB, Cube RV, Hutchinson JH, James JK, Gardner MF, Schaffhauser H, Rowe BA, Daggett LP, Vernier JM (2004) Allosteric potentiators of the metabotropic glutamate receptor 2 (mGlu2): Part 2. 4-thiopyridyl acetophenones as non-tetrazole containing mGlu2 receptor potentiators. Bioorg Med Chem Lett 14:5867–5872

66. Pinkerton AB, Cube RV, Hutchinson JH, Rowe BA, Schaffhauser H, Zhao X, Daggett LP, Vernier JM (2004) Allosteric potentiators of the metabotropic glutamate receptor 2 (mGlu2): Part 1. Identification and synthesis of phenyl-tetrazolyl acetophenones. Bioorg Med Chem Lett 14:5329–5332

67. Benneyworth MA, Xiang Z, Smith RL, Garcia EE, Conn PJ, Sanders-Bush E (2007) A selective positive allosteric modulator of metabotropic glutamate receptor subtype 2 blocks a hallucinogenic drug model of psychosis. Mol Pharmacol 72:477–484

68. Galici R, Jones CK, Hemstapat K, Nong Y, Echemendia NG, Williams LC, de Paulis T, Conn PJ (2006) Biphenyl-indanone A, a positive allosteric modulator of the metabotropic glutamate receptor subtype 2, has antipsychotic- and anxiolytic-like effects in mice. J Pharmacol Exp Ther 318:173–185

69. Phillips AG, Fibiger HC (1990) Role of reward and enhancement of conditioned reward in persistence of responding for cocaine. Behav Pharmacol 1:269–282

70. Rice ME, Cragg SJ (2004) Nicotine amplifies reward-related dopamine signals in striatum. Nat Neurosci 7:583–584

71. Robbins TW, Watson BA, Gaskin M, Ennis C (1983) Contrasting interactions of pipradrol, d-amphetamine, cocaine, cocaine analogues, apomorphine and other drugs with conditioned reinforcement. Psychopharmacology (Berl) 80:113–119

72. Taylor JR, Robbins TW (1986) 6-Hydroxydopamine lesions of the nucleus accumbens, but not of the caudate nucleus, attenuate enhanced responding with reward-related stimuli produced by intra-accumbens d-amphetamine. Psychopharmacology (Berl) 90:390–397

73. Chaudhri N, Caggiula AR, Donny EC, Booth S, Gharib M, Craven L, Palmatier MI, Liu X, Sved AF (2006) Operant responding for conditioned and unconditioned reinforcers in rats is differentially enhanced by the primary reinforcing and reinforcement-enhancing effects of nicotine. Psychopharmacology (Berl) 189:27–36

74. Kenny PJ (2007) Brain reward systems and compulsive drug use. Trends Pharmacol Sci 28:135–141

75. Kenny PJ, Markou A (2006) Nicotine self-administration acutely activates brain reward systems and induces a long-lasting increase in reward sensitivity. Neuropsychopharmacology 31:1203–1211

76. Harrison AA, Gasparini F, Markou A (2002) Nicotine potentiation of brain stimulation reward reversed by DHβE and SCH 23390, but not by eticlopride, LY 314582 or MPEP in rats. Psychopharmacology (Berl) 160:56–66

77. Markou A, Koob GF (1993) Intracranial self-stimulation thresholds are a measure of reward. In: Saghal A (ed) Behavioural neuroscience: a practical approach, vol 2. IRL Press, Oxford, pp 93–115

78. Vlachou S, Markou A (2010) Intracranial self-stimulation: the use of the intracranial self-stimulation procedure in the investigation of reward and motivational processes: effects of drugs of abuse. In: Olmstead MC (ed) Animal models of drug addiction, Humana Press, New York (in press)

79. Kenny PJ, Gasparini F, Markou A (2003) Group II metabotropic and α-amino-3-hydroxy-5-methyl-4-isoxazole propionate (AMPA)/kainate glutamate receptors regulate the deficit in brain reward function associated with nicotine withdrawal in rats. J Pharmacol Exp Ther 306:1068–1076

80. Sanchis-Segura C, Spanagel R (2006) Behavioural assessment of drug reinforcement and addictive features in rodents: an overview. Addict Biol 11:2–38

81. McGeehan AJ, Olive MF (2003) The mGluR5 antagonist MPEP reduces the conditioned rewarding effects of cocaine but not other drugs of abuse. Synapse 47:240–242

82. Miyatake M, Narita M, Shibasaki M, Nakamura A, Suzuki T (2005) Glutamatergic neuro-transmission and protein kinase C play a role in neuron-glia communication during the development of methamphetamine-induced psychological dependence. Eur J Neurosci 22:1476–1488

83. Herzig V, Capuani EM, Kovar KA, Schmidt WJ (2005) Effects of MPEP on expression of food-, MDMA- or amphetamine-conditioned place preference in rats. Addict Biol 10:243–249

84. van der Kam EL, De Vry J, Tzschentke TM (2009) 2-Methyl-6-(phenylethynyl)-pyridine (MPEP) potentiates ketamine and heroin reward as assessed by acquisition, extinction, and reinstatement of conditioned place preference in the rat. Eur J Pharmacol 606:94–101

85. Aoki T, Narita M, Shibasaki M, Suzuki T (2004) Metabotropic glutamate receptor 5 localized in the limbic forebrain is critical for the development of morphine-induced rewarding effect in mice. Eur J Neurosci 20:1633–1638

86. Popik P, Wrobel M (2002) Morphine conditioned reward is inhibited by MPEP, the mGluR5 antagonist. Neuropharmacology 43:1210–1217

87. Herzig V, Schmidt WJ (2004) Effects of MPEP on locomotion, sensitization and conditioned reward induced by cocaine or morphine. Neuropharmacology 47:973–984

88. Varty GB, Grilli M, Forlani A, Fredduzzi S, Grzelak ME, Guthrie DH, Hodgson RA, Lu SX, Nicolussi E, Pond AJ et al (2005) The antinociceptive and anxiolytic-like effects of the metabotropic glutamate receptor 5 (mGluR5) antagonists, MPEP and MTEP, and the mGluR1 antagonist, LY456236, in rodents: a comparison of efficacy and side-effect profiles. Psychopharmacology (Berl) 179:207–217

89. Nakagawa T, Fujio M, Ozawa T, Minami M, Satoh M (2005) Effect of MS-153, a glutamate transporter activator, on the conditioned rewarding effects of morphine, methamphetamine and cocaine in mice. Behav Brain Res 156:233–239

90. Popik P, Kozela E, Wrobel M, Wozniak KM, Slusher BS (2003) Morphine tolerance and reward but not expression of morphine dependence are inhibited by the selective glutamate carboxypeptidase II (GCP II, NAALADase) inhibitor, 2-PMPA. Neuropsychopharmacology 28:457–467

91. Slusher BS, Thomas A, Paul M, Schad CA, Ashby CR Jr (2001) Expression and acquisition of the conditioned place preference response to cocaine in rats is blocked by selective inhibitors of the enzyme N-acetylated-α-linked-acidic dipeptidase (NAALADASE). Synapse 41:22–28

92. Malin DH, Lake JR, Newlin-Maultsby P, Roberts LK, Lanier JG, Carter VA, Cunningham JS, Wilson OB (1992) Rodent model of nicotine abstinence syndrome. Pharmacol Biochem Behav 43:779–784

93. Epping-Jordan MP, Watkins SS, Koob GF, Markou A (1998) Dramatic decreases in brain reward function during nicotine withdrawal. Nature 393:76–79

94. Markou A, Kosten TR, Koob GF (1998) Neurobiological similarities in depression and drug dependence: a self-medication hypothesis. Neuropsychopharmacology 18:135–174

95. Kenny PJ, Markou A (2004) The ups and downs of addiction: role of metabotropic glutamate receptors. Trends Pharmacol Sci 25:265–272

96. Epstein DH, Preston KL, Stewart J, Shaham Y (2006) Toward a model of drug relapse: an assessment of the validity of the reinstatement procedure. Psychopharmacology (Berl) 189:1–16

97. Reichel CM, Bevins RA (2009) Forced abstinence model of relapse to study pharmacologi-cal treatments of substance use disorder. Curr Drug Abuse Rev 2:184–194

98. Shiffman S, Balabanis MH, Gwaltney CJ, Paty JA, Gnys M, Kassel JD, Hickcox M, Paton SM (2007) Prediction of lapse from associations between smoking and situational antecedents assessed by ecological momentary assessment. Drug Alcohol Depend 91:159–168

99. Shiffman S, Waters AJ (2004) Negative affect and smoking lapses: a prospective analysis. J Consult Clin Psychol 72:192–201

100. Baker TB, Piper ME, McCarthy DE, Majeskie MR, Fiore MC (2004) Addiction motivation reformulated: an affective processing model of negative reinforcement. Psychol Rev 111:33–51

101. Abrantes AM, Strong DR, Lejuez CW, Kahler CW, Carpenter LL, Price LH, Niaura R, Brown RA (2008) The role of negative affect in risk for early lapse among low distress tolerance smokers. Addict Behav 33:1394–1401

102. Brown RA, Lejuez CW, Strong DR, Kahler CW, Zvolensky MJ, Carpenter LL, Niaura R, Price LH (2009) A prospective examination of distress tolerance and early smoking lapse in adult self-quitters. Nicotine Tob Res 11:493–502

103. Markou A, Koob GF (1991) Postcocaine anhedonia: an animal model of cocaine withdrawal. Neuropsychopharmacology 4:17–26

104. Paterson NE, Myers C, Markou A (2000) Effects of repeated withdrawal from continuous amphetamine administration on brain reward function in rats. Psychopharmacology (Berl) 152:440–446

105. Lin D, Koob GF, Markou A (2000) Time-dependent alterations in ICSS thresholds associated with repeated amphetamine administrations. Pharmacol Biochem Behav 65:407–417

106. Schulteis G, Markou A, Gold LH, Stinus L, Koob GF (1994) Relative sensitivity to naloxone of multiple indices of opiate withdrawal: a quantitative dose-response analysis. J Pharmacol Exp Ther 271:1391–1398

107. Schulteis G, Markou A, Cole M, Koob GF (1995) Decreased brain reward produced by ethanol withdrawal. Proc Natl Acad Sci USA 92:5880–5884

108. Watkins SS, Stinus L, Koob GF, Markou A (2000) Reward and somatic changes during precipitated nicotine withdrawal in rats: centrally and peripherally mediated effects. J Pharmacol Exp Ther 292:1053–1064

109. Watkins SS, Koob GF, Markou A (2000) Neural mechanisms underlying nicotine addiction: acute positive reinforcement and withdrawal. Nicotine Tob Res 2:19–37

110. Sepulveda J, Oliva P, Contreras E (2004) Neurochemical changes of the extracellular concentrations of glutamate and aspartate in the nucleus accumbens of rats after chronic administration of morphine. Eur J Pharmacol 483:249–258

111. Palucha A, Branski P, Pilc A (2004) Selective mGlu5 receptor antagonist MTEP attenuates naloxone-induced morphine withdrawal symptoms. Pol J Pharmacol 56:863–866

112. Rasmussen K, Martin H, Berger JE, Seager MA (2005) The mGlu5 receptor antagonists MPEP and MTEP attenuate behavioral signs of morphine withdrawal and morphine-withdrawal-induced activation of locus coeruleus neurons in rats. Neuropharmacology 48:173–180

113. Fundytus ME, Coderre TJ (1997) Attenuation of precipitated morphine withdrawal symptoms by acute i.c.v. administration of a group II mGluR agonist. Br J Pharmacol 121:511–514

114. Vandergriff J, Rasmussen K (1999) The selective mGlu2/3 receptor agonist LY354740 attenuates morphine-withdrawal-induced activation of locus coeruleus neurons and behavioral signs of morphine withdrawal. Neuropharmacology 38:217–222

115. Klodzinska A, Chojnacka-Wojcik E, Palucha A, Branski P, Popik P, Pilc A (1999) Potential anti-anxiety, anti-addictive effects of LY 354740, a selective group II glutamate metabotropic receptors agonist in animal models. Neuropharmacology 38:1831–1839

116. Liechti ME, Markou A (2007) Metabotropic glutamate 2/3 receptor activation induced reward deficits but did not aggravate brain reward deficits associated with spontaneous nicotine withdrawal in rats. Biochem Pharmacol 74:1299–1307

117. Fundytus ME, Ritchie J, Coderre TJ (1997) Attenuation of morphine withdrawal symptoms by subtype-selective metabotropic glutamate receptor antagonists. Br J Pharmacol 120:1015–1020

118. Rasmussen K, Hsu MA, Vandergriff J (2004) The selective mGlu2/3 receptor antagonist LY341495 exacerbates behavioral signs of morphine withdrawal and morphine-withdrawal-induced activation of locus coeruleus neurons. Neuropharmacology 46:620–628

119. Chaki S, Yoshikawa R, Hirota S, Shimazaki T, Maeda M, Kawashima N, Yoshimizu T, Yasuhara A, Sakagami K, Okuyama S et al (2004) MGS0039: a potent and selective group II

metabotropic glutamate receptor antagonist with antidepressant-like activity. Neuropharmacology 46:457–467

120. Karasawa J, Yoshimizu T, Chaki S (2006) A metabotropic glutamate 2/3 receptor antagonist, MGS0039, increases extracellular dopamine levels in the nucleus accumbens shell. Neurosci Lett 393:127–130

121. Barr AM, Markou A (2005) Psychostimulant withdrawal as an inducing condition in animal models of depression. Neurosci Biobehav Rev 29:675–706

122. Barr AM, Markou A, Phillips AG (2002) A "crash" course on psychostimulant withdrawal as a model of depression. Trends Pharmacol Sci 23:475–482

123. Paterson NE, Markou A (2007) Animal models and treatments for addiction and depression co-morbidity. Neurotox Res 11:1–32

124. Cryan JF, Markou A, Lucki I (2002) Assessing antidepressant activity in rodents: recent developments and future needs. Trends Pharmacol Sci 23:238–245

125. Bruijnzeel AW, Zislis G, Wilson C, Gold MS (2007) Antagonism of CRF receptors prevents the deficit in brain reward function associated with precipitated nicotine withdrawal in rats. Neuropsychopharmacology 32:955–963

126. Sarnyai Z, Biro E, Gardi J, Vecsernyes M, Julesz J, Telegdy G (1995) Brain corticotropin-releasing factor mediates "anxiety-like" behavior induced by cocaine withdrawal in rats. Brain Res 675:89–97

127. Basso AM, Spina M, Rivier J, Vale W, Koob GF (1999) Corticotropin-releasing factor antagonist attenuates the "anxiogenic-like" effect in the defensive burying paradigm but not in the elevated plus-maze following chronic cocaine in rats. Psychopharmacology (Berl) 145:21–30

128. Schulteis G, Yackey M, Risbrough V, Koob GF (1998) Anxiogenic-like effects of spontaneous and naloxone-precipitated opiate withdrawal in the elevated plus-maze. Pharmacol Biochem Behav 60:727–731

129. Harris GC, Aston-Jones G (1993) β-adrenergic antagonists attenuate withdrawal anxiety in cocaine- and morphine-dependent rats. Psychopharmacology (Berl) 113:131–136

130. Knapp DJ, Overstreet DH, Moy SS, Breese GR (2004) SB242084, flumazenil, and CRA1000 block ethanol withdrawal-induced anxiety in rats. Alcohol 32:101–111

131. Overstreet DH, Knapp DJ, Breese GR (2004) Similar anxiety-like responses in male and female rats exposed to repeated withdrawals from ethanol. Pharmacol Biochem Behav 78:459–464

132. Funk CK, Koob GF (2007) A CRF_2 agonist administered into the central nucleus of the amygdala decreases ethanol self-administration in ethanol-dependent rats. Brain Res 1155:172–178

133. George O, Ghozland S, Azar MR, Cottone P, Zorrilla EP, Parsons LH, O'Dell LE, Richardson HN, Koob GF (2007) CRF-CRF_1 system activation mediates withdrawal-induced increases in nicotine self-administration in nicotine-dependent rats. Proc Natl Acad Sci USA 104:17198–17203

134. Jonkman S, Risbrough VB, Geyer MA, Markou A (2008) Spontaneous nicotine withdrawal potentiates the effects of stress in rats. Neuropsychopharmacology 33:2131–2138

135. Engelmann JM, Radke AK, Gewirtz JC (2009) Potentiated startle as a measure of the negative affective consequences of repeated exposure to nicotine in rats. Psychopharmacology (Berl) 207:13–25

136. Davis M, Falls WA, Campeau S, Kim M (1993) Fear-potentiated startle: a neural and pharmacological analysis. Behav Brain Res 58:175–198

137. Grillon C, Cordova J, Levine LR, Morgan CA 3rd (2003) Anxiolytic effects of a novel group II metabotropic glutamate receptor agonist (LY354740) in the fear-potentiated startle paradigm in humans. Psychopharmacology (Berl) 168:446–454

138. Dunayevich E, Erickson J, Levine L, Landbloom R, Schoepp DD, Tollefson GD (2008) Efficacy and tolerability of an mGlu2/3 agonist in the treatment of generalized anxiety disorder. Neuropsychopharmacology 33:1603–1610

139. Grillon C, Avenevoli S, Daurignac E, Merikangas KR (2007) Fear-potentiated startle to threat, and prepulse inhibition among young adult nonsmokers, abstinent smokers, and nonabstinent smokers. Biol Psychiatry 62:1155–1161

140. Weiss F (2005) Neurobiology of craving, conditioned reward and relapse. Curr Opin Pharmacol 5:9–19

141. Dackis CA, O'Brien CP (2001) Cocaine dependence: a disease of the brain's reward centers. J Subst Abuse Treat 21:111–117

142. O'Brien CP, Childress AR, Ehrman R, Robbins SJ (1998) Conditioning factors in drug abuse: can they explain compulsion? J Psychopharmacol 12:15–22

143. O'Brien CP, McLellan AT (1996) Myths about the treatment of addiction. Lancet 347:237–240

144. Backstrom P, Hyytia P (2006) Ionotropic and metabotropic glutamate receptor antagonism attenuates cue-induced cocaine seeking. Neuropsychopharmacology 31:778–786

145. Bespalov AY, Dravolina OA, Sukhanov I, Zakharova E, Blokhina E, Zvartau E, Danysz W, van Heeke G, Markou A (2005) Metabotropic glutamate receptor (mGluR5) antagonist MPEP attenuated cue- and schedule-induced reinstatement of nicotine self-administration behavior in rats. Neuropharmacology 49(Suppl 1):167–178

146. Backstrom P, Bachteler D, Koch S, Hyytia P, Spanagel R (2004) mGluR5 antagonist MPEP reduces ethanol-seeking and relapse behavior. Neuropsychopharmacology 29:921–928

147. Dravolina OA, Zakharova ES, Shekunova EV, Zvartau EE, Danysz W, Bespalov AY (2007) mGlu1 receptor blockade attenuates cue- and nicotine-induced reinstatement of extinguished nicotine self-administration behavior in rats. Neuropharmacology 52:263–269

148. Lu L, Uejima JL, Gray SM, Bossert JM, Shaham Y (2007) Systemic and central amygdala injections of the mGluR$_{2/3}$ agonist LY379268 attenuate the expression of incubation of cocaine craving. Biol Psychiatry 61:591–598

149. Peters J, Kalivas PW (2006) The group II metabotropic glutamate receptor agonist, LY379268, inhibits both cocaine- and food-seeking behavior in rats. Psychopharmacology (Berl) 186:143–149

150. Bossert JM, Gray SM, Lu L, Shaham Y (2006) Activation of group II metabotropic glutamate receptors in the nucleus accumbens shell attenuates context-induced relapse to heroin seeking. Neuropsychopharmacology 31:2197–2209

151. Bossert JM, Liu SY, Lu L, Shaham Y (2004) A role of ventral tegmental area glutamate in contextual cue-induced relapse to heroin seeking. J Neurosci 24:10726–10730

152. Zhao Y, Dayas CV, Aujla H, Baptista MA, Martin-Fardon R, Weiss F (2006) Activation of group II metabotropic glutamate receptors attenuates both stress and cue-induced ethanol-seeking and modulates c-*fos* expression in the hippocampus and amygdala. J Neurosci 26:9967–9974

153. Cartmell J, Monn JA, Schoepp DD (1999) The metabotropic glutamate 2/3 receptor agonists LY354740 and LY379268 selectively attenuate phencyclidine versus d-amphetamine motor behaviors in rats. J Pharmacol Exp Ther 291:161–170

154. Cartmell J, Monn JA, Schoepp DD (2000) Attenuation of specific PCP-evoked behaviors by the potent mGlu2/3 receptor agonist, LY379268 and comparison with the atypical antipsychotic, clozapine. Psychopharmacology (Berl) 148:423–429

155. Moran MM, McFarland K, Melendez RI, Kalivas PW, Seamans JK (2005) Cystine/glutamate exchange regulates metabotropic glutamate receptor presynaptic inhibition of excitatory transmission and vulnerability to cocaine seeking. J Neurosci 25:6389–6393

156. Baker DA, McFarland K, Lake RW, Shen H, Tang XC, Toda S, Kalivas PW (2003) Neuroadaptations in cystine-glutamate exchange underlie cocaine relapse. Nat Neurosci 6:743–749

157. Madayag A, Lobner D, Kau KS, Mantsch JR, Abdulhameed O, Hearing M, Grier MD, Baker DA (2007) Repeated *N*-acetylcysteine administration alters plasticity-dependent effects of cocaine. J Neurosci 27:13968–13976

158. Zhou W, Kalivas PW (2008) N-acetylcysteine reduces extinction responding and induces enduring reductions in cue- and heroin-induced drug-seeking. Biol Psychiatry 63:338–340

159. LaRowe SD, Mardikian P, Malcolm R, Myrick H, Kalivas P, McFarland K, Saladin M, McRae A, Brady K (2006) Safety and tolerability of N-acetylcysteine in cocaine-dependent individuals. Am J Addict 15:105–110

160. LaRowe SD, Myrick H, Hedden S, Mardikian P, Saladin M, McRae A, Brady K, Kalivas PW, Malcolm R (2007) Is cocaine desire reduced by N-acetylcysteine? Am J Psychiatry 164:1115–1117

161. Mardikian PN, LaRowe SD, Hedden S, Kalivas PW, Malcolm RJ (2007) An open-label trial of N-acetylcysteine for the treatment of cocaine dependence: a pilot study. Prog Neuropsychopharmacol Biol Psychiatry 31:389–394

162. Grant JE, Kim SW, Odlaug BL (2007) N-acetyl cysteine, a glutamate-modulating agent, in the treatment of pathological gambling: a pilot study. Biol Psychiatry 62:652–657

163. Knackstedt LA, LaRowe S, Mardikian P, Malcolm R, Upadhyaya H, Hedden S, Markou A, Kalivas PW (2009) The role of cystine-glutamate exchange in nicotine dependence in rats and humans. Biol Psychiatry 65:841–845

164. Ballard TM, Woolley ML, Prinssen E, Huwyler J, Porter R, Spooren W (2005) The effect of the mGlu5 receptor antagonist MPEP in rodent tests of anxiety and cognition: a comparison. Psychopharmacology (Berl) 179:218–229

Metabotropic Approaches to Anxiety

Joanna M. Wieronska, Gabriel Nowak, and Andrzej Pilc

Abstract Current therapies for anxiety disorders neither fully serve the efficacy needs of patients nor are they free of adverse effects. Both preclinical and clinical findings have implicated the excitatory amino acid glutamate in the pathogenesis of anxiety disorders. While a number of review papers were published in recent years describing the anxiolytic effect of mGlu receptor ligands, in this short review we try to explain the mechanisms responsible for the antianxiety actions of specific mGlu receptors ligands, which have been reported to be potent anxiolytics. As the amygdala is the structure integrating the behaviors connected with fear and anxiety, the schema of amygdalar nuclei is shown together with the placement of mGu receptors in that structure. Furthermore, the anxiolytic effect of different mGlu receptor ligands in the context of their activity within specific amygdalar nuclei is proposed. The interactions between the anxiolytic effects of mGlu receptor ligands and other neurotransmitters involved in anxiety, particularly GABA and serotonin, will also be discussed, and the neuronal networks involved will be described in order to discuss the mechanism of the proposed anxiolytic effects.

1 Introduction

Anxiety disorders are a group of mental diseases affecting the greatest number of people both in the United States and in European countries. Medical reports inform that in the EU countries over 41 million people suffer from anxiety disorders [1] and in the USA the number of patients is estimated at 20–26 million [2, 3]. The costs of treating the disease are very high and reach nearly $42 billion per year in the EU [1].

A. Pilc (✉)
Institute of Pharmacology PAS, Smetna 12, 31-343 Kraków, Poland
Collegium Medicum, Jagiellonian University, 31-531 Kraków, Poland
e-mail: nfpilc@cyf-kr.edu.pl

P. Skolnick (ed.), *Glutamate-based Therapies for Psychiatric Disorders*,
Milestones in Drug Therapy, DOI 10.1007/978-3-0346-0241-9_9,
© Springer Basel AG 2010

Current psychiatric diagnostic criteria includes a wide variety of anxiety disorders such as generalized anxiety disorder, phobias, panic attacks, posttraumatic stress disorder, and several others.

Despite this heterogeneity in the overall behavioral picture of anxiety symptoms, it is speculated that the increased activity of glutamate, the major excitatory amino acid in the brain, may be essential in the pathophysiology of anxious states [4, 5]. The important role of GABAergic inhibitory control over the glutamatergic network should also be taken into consideration, as the balance between these two major amino acid neurotransmitters in the central nervous system (CNS) ensures its proper functioning [6]. The present pharmacotherapy of anxiety is based on increasing GABAergic activity through the agonistic action on the $GABA_A$ receptors [7] or the use of antidepressant drugs, particularly serotonin reuptake inhibitors [8]. The final result of both these types of treatment may lead to an indirect decrease of the glutamatergic system activity, which was partially established via electrophysiological studies [9] as well as by the in vivo microdialysis [10, 11]. Therefore, the direct inhibition of increased glutamatergic activity in anxiety may be a new target, with metabotropic glutamate (mGlu) receptors as the main site of action of novel anxiolytics.

Limited clinical proof of concept for mGlu receptor ligands in the treatment of anxiety disorders has been achieved. The mGlu2/3 receptor agonist, LY544344, has shown efficacy in the treatment of generalized anxiety disorder in a double blind, placebo-controlled study [12]. Fenobam, a nonbenzodiazepine anxiolytic in humans, may produce this clinical outcome through the blockade of specific metabotropic glutamate receptor (mGlu5) [13]. Nonetheless, support for the role of mGlu receptors in the anxiety disorders presented here rests primarily on the weight of preclinical evidence provided by the limited information on these ligands in humans.

Below, the main neuronal circuit involved in anxiety as well as the classification and distribution of the mGlu receptors in the CNS and the anxiolytic activity of their ligands in animal models will be described. Moreover, the putative mechanism of action based on the localization of mGlu receptors in the synapses within key brain structures involved in the regulation of fear and anxiety will also be introduced. The interactions between the anxiolytic effects of the mGlu receptor ligands and other neurotransmitters, particularly involved in anxiety, mainly GABA and serotonin, will be also discussed.

2 The Circuit of Fear and Anxiety

A complex response to stress is a consequence of the cascade of events, such as receiving, integrating and consolidating the information of fear. During this process, the hypothalamus, activated by external stimuli, is supposed to integrate the diverse and dynamic environmental cues with the biological requirements of the individual [14, 15]. Hypothalamic projections to the structures with the well-documented key role in emotional response, such as amygdala (Amy) and periaqueductal gray (PAG), are supposed to be the major network regulating anxiety behavior. The control of emotional responses mediated by the amygdala or PAG is maintained by the afferent connections with cortical and subcortical brain areas, such as the hippocampus.

It is well established that the amygdala plays a pivotal role in fear. The findings of functional neuroimaging studies show exaggerated amygdala activation to specific stimuli in a number of anxiety disorders [14], which are then normalized by medication (antidepressants or benzodiazepines) [16, 17]. The amygdaloid complex constitutes of several nuclei which are grouped into larger compartments and together form an almond-shape structure of the limbic system, located in the medial temporal lobe in the mammalian brain, including humans. In general the amygdalar nuclei are divided into cortical, basolateral (which includes lateral, basolateral, ventral basolateral and basomedial nuclei), medial and central [18, 19]. A schematic diagram of the amygdala section is shown in Fig. 1.

The main glutamatergic thalamo–cortical input to the amygdala is received by its lateral nucleus. The information is transferred through the network of pyramidal neurons to the other amygdalar nuclei, where it is then processed and transformed.

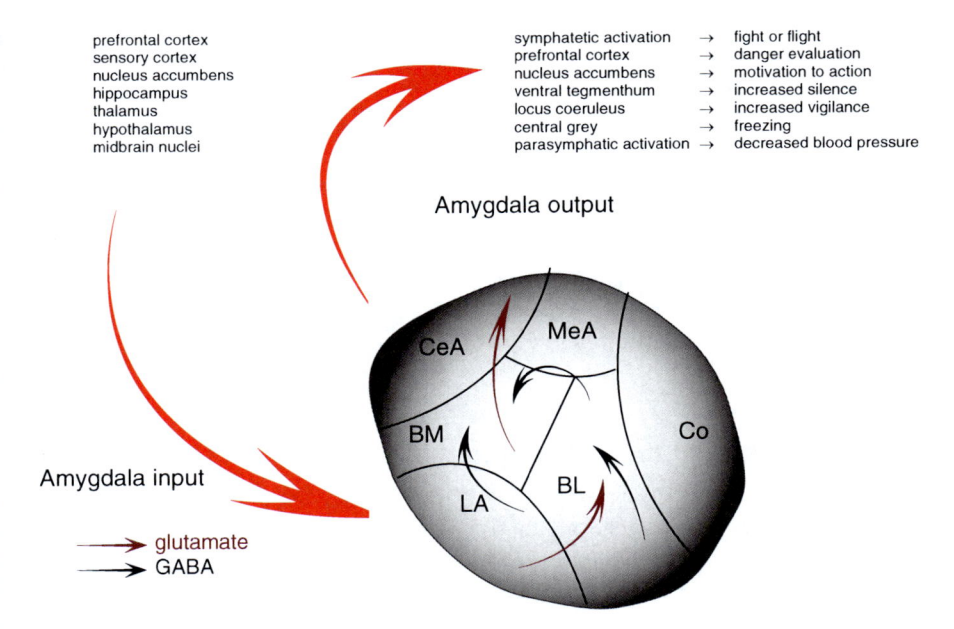

Fig. 1 A schematic representation of the amygdala: *LA* lateral nucleus; *BL* basolateral nucleus; *BM* basomedial nucleus; *Co* cortical nucleus; *MeA* medial nucleus; *CeA* central nucleus. The *red arrows* represent the main glutamatergic input and output of the amygdala, as well as the pyramidal neuronal network within the nucleus. The *black arrows* represent the amygdala interneurons controlling glutamate activity

The LA is the locus where the information from other areas is received. The signal is sent to the CeA and after its activation the behavioral, autonomic and endocrine components of the stress response are exerted by activating particular areas of the brain

The central nucleus is the final point of information processing and the main amygdala output [20]. The glutamatergic efferents innervate executive structures and evoke responses on the physiological, behavioral, and vegetative levels characteristic of anxiety [18, 19]. The network of glutamatergic pyramidal neurons is controlled by GABAergic inhibitory interneurons, located within the nucleus (Fig. 1). In the physiological state, all mechanisms are properly functioning and the balance between glutamate and GABA maintains emotional responses at the level appropriate to external stimuli. An upset of the balance leads to disturbances in the neurotransmitters' levels and in a state of anxiety the lack of GABAergic control over the glutamatergic system is proposed, which leads to increased excitation in the brain (Fig. 1) [21]. Overactivity of the amygdala can be a result of (1) increased firing of the amygdala glutamatergic input or (2) disinhibition of the pyramidal network within amygdalar nuclei, leading to overexcitation of efferent projections. One of the structures receiving the information from the amygdala is the PAG, where the dorsal and dorsolateral parts are especially involved in adjusting the somatic and neurovegetative elements in the defensive response.

In higher vertebrates, including humans, the information transferred from hypothalamus to either the amygdala or PAG is evaluated in the prefrontal cortex (PFC) and hippocampus, regulating the conscious reaction to emotional stimuli, imperative to the extinction and renewal of fear response [15]. During extinction, inputs from the PFC activate extinction neurons in the basolateral amygdala, which then inhibit the activity of fear output neurons in the central nucleus of the amygdala (CeA). During fear renewal, inputs from the hippocampus activate inhibitory interneurons in the basolateral amygdala that silence extinction neurons restoring the stress response [15]. The importance of hippocampal formation in the anxiolytic action of mGlu receptor ligands is well documented as a variety of studies indicated that their intrahippocampal administration evoked such activity (for discussion see [22]). The most probable mechanism of anxiolytic action of those ligands is in diminishing the activity of efferent glutamatergic neurons.

3   Distribution of mGlu Receptors in the Synaptic Cleft

The distribution of particular mGlu receptors subtypes within the brain areas associated with fear and anxiety differs, and this has an impact on the regulation of neurotransmission exerted by their ligands. Their localization in the synaptic cleft can be both post- and presynaptic, depending on the receptor subtype. The postsynaptic localization, described mainly for the group I mGlu receptors, allows for mediating a slow excitatory current after ligand binding [23]. Activation of presynaptic receptors has an impact on neurotransmitter release (e.g., as auto or heteroreceptors they regulate either the release of glutamate or other neurotransmitter). Presynaptic localization is predominant for the group II and III mGlu receptors, although the postsynaptic localization of these receptors was also described. The affinity of the natural agonists at particular types of the receptors differs, and seems to conform with their expression in the synaptic cleft [23, 24]. Group II receptors, mGlu2 and mGlu3, were shown to be localized not in close association with the synapse, but rather at the extrasynaptic dendritic membrane, often along axon terminals at various distances from the synaptic junction [25]. Their affinity to the glutamate is the highest of all mGlu receptor subtypes, and varies from 0.3 to 12 μM.

Among all mGlu receptors, the most privileged localization was described for the mGlu7 subtype, which is expressed commonly around synaptic vesicles or in the presynaptic active zones in axon terminals. The affinity of L-glutamate for this receptor is the lowest among all mGlu receptors, and millimolar glutamate concentrations are needed for its activation [26]. Electron microscope studies revealed that the receptor can be localized on pyramidal neurons, thus inhibiting glutamate release, but its exclusive distribution was shown on GABAergic terminals innervating interneurons [27, 28]. Such localization was also observed for other members of the mGlu group III receptor family.

4 Amygdala and mGlu Receptors

Studies concerning the distribution of particular mGlu receptor subtypes in the amygdala revealed that their localization is not homogenous. A relatively weak expression was described for mGlu1 and mGlu3 subtypes, restricted mainly to basolateral and central nuclei [29], and for the mGlu8 receptor detected mainly in the lateral nucleus [30]. Therefore, the absence or weak anxiolytic action of compounds acting at these receptors is not surprising [31].

Lateral amygdala, the nuclei of the main amygdala input, was shown to be enriched in mGlu5, mGlu2/3, and mGlu4 receptors [29, 32, 33] and ligands of these receptors were shown to possess excellent anxiolytic activity in animal models [21, 22]. Similarly, such activity was observed for the mGlu7 receptor allosteric modulator, AMN082, which activates mGlu7 receptors, located in relatively high density in both cortical and central nuclei [34]. A summarization of the distribution of mGlu receptors in amygdalar nuclei is shown in Table 1.

Table 1 Distribution of immunoreactivity for mGlu receptors within the amygdala. Immunoreactivity: +++ very strong; ++ strong; + moderate to weak; − very weak, if any; ? not tested

	Amygdala nuclei			
	Lateral	Basolateral	Central	Cortical
mGlu1	−	+	+	−
mGlu5	++	+	+	+
mGlu2/3	+++	++	+	−
mGlu3	−	−	+	−
mGlu4	++	++	−	−
mGlu7	−	−	++	+++
mGlu8	++	?	?	?

5 Mechanism of Action of mGlu Ligands

A comparison of mGlu receptors ligands with anxiolytic activity can be found in many review papers [21, 22, 35]. Analyzing the number of papers published, it can be concluded that the majority of positive results concern the anxiolytic activity of groups I and II ligands, with 162 and 99 reports published, respectively, since 1996. The studies regarding the group III are more limited because of the restricted number of subtype-specific and brain-penetrating agents, although 50 reports on the subject have been published since 1996; for reviews see [21, 22, 26, 35, 36].

Allosteric modulation seems to be a very promising way to influence glutamate activity. The discovery of allosteric modulators of mGlu receptors represents an exciting advance in demonstrating the potential for developing novel research tools and therapeutic agents to regulate the activity of specific mGlu receptors subtypes. Negative allosteric modulators of mGlu5 receptors and positive modulators of mGlu2 receptors are the best candidates for novel anxiolytics [37–41]. Ligands of group III mGlu receptors are still undergoing close examination.

An understanding of the mechanism of action of anxiolytics is informed by the neuronal basis of fear and anxiety. The behavioral tasks used to assess the anxiety in animals fall into learned and unlearned. Unlearned fear relies on stimuli not associated with any experience, and learned fear is provoked by the conditioned stimuli associated with an aversive situation. In some paradigms, the autonomic response to stress (such as a decrease in body temperature) is also taken into consideration [42]. The effectiveness of mGlu ligands was shown in all types of anxiety models. The fundamental mechanism of their anxiolytic action may be related to restoring the increased glutamatergic cascade of events within the amygdala during a pathological state. Below, there is an attempt to summarize the results in the light of the putative neuronal basics of mGlu-mediated anxiolysis in different types of fear.

One of the major animal models for investigating the neuronal basis for emotional learning is fear conditioning (CF), measured by the fear-potentiated startle (FPS) paradigm and freezing. The basis of the paradigm is the association of the neutral stimulus with an aversive, unconditioned stimulus (US). After a few days of training, the state of fear is elicited only by the occurrence of initially neutral, conditioned stimulus (CS). The training procedure reflects the acquisition, consolidation, and expression of CF and the amygdala, particularly its lateral nucleus is the point where the conditional and unconditional stimuli converge to produce fear conditioning [43]. The long-term potentiation (LTP) of synapses from thalamic/hypothalamic afferents to LA neurons is critical to this process [44, 45] and stimulation of this afferent pathway leads to CF. An important role of NMDA receptor and non-NMDA glutamate receptors was reported [46]. Recently, multiple behavioral studies have revealed that metabotropic glutamate receptors are strongly involved in regulation of LTP as well, by influencing the acquisition, consolidation, or expression of CF. Multiple studies have shown that administration of mGlu receptors ligands influenced the behavioral events connected with fear conditioning and the induction/inhibition of the LTP paradigm.

5.1 Group I mGlu Receptors

Numerous studies demonstrated the robust effects of negative allosteric modulators of mGlu5 receptors in a variety of animal tests used to detect the anxiolytic action of psychotropic compounds (for reviews see [36, 47, 48]). The mGlu5 receptor negative modulator, MPEP, was also effective in the abolishing the acquisition, but not the consolidation and expression of CF when injected into LA [49]. These behavioral studies were followed by whole cell patch-clamp recordings showing inhibition of amygdaloid LTP in the CS pathway [49].

The mechanism of action of mGlu5 receptor negative modulators seems to be directly dependent on serotonergic, but not GABAergic neurotransmission mediated through GABA receptor signaling, at least when assessed in Vogel's conflict test or in the plus-maze paradigm [29, 50, 51]. The serotonergic 5-HT2A receptor blocker, ritanserin, abolished MTEP-mediated anxiolysis while flumazenil, the $GABA_A$ receptor blocker, was ineffective [50]. Reflecting upon the mechanism of action of MTEP in the light of these results, a cascade of consecutive events must be taken into consideration, which would result in a decrease of glutamatergic neurotransmission at the final step. The schematic representation of the sequential signal transfer from the amygdala input to efferent connections is shown in Fig. 2. In brief, the excess of glutamate released by the presynaptic pyramidal afferent neuron acts on the postsynaptic pyramidal element leading to the excitation of the amygdalar glutamatergic network. Released glutamate acts through mGlu5 receptors linked by Homer proteins to NMDA receptors localized postsynaptically. Negative modulation of mGlu5 receptors exerted by MTEP (which binds to the allosteric binding site within the transmembrane domain) leads to a lower affinity of glutamate to its binding site on the mGlu5 receptor, which results in an inhibition (e.g., normalization) of glutamatergic system activity.

The inhibition of the anxiolytic action of MTEP by serotonergic agents synonymously indicates that the serotonergic system is involved in its effectiveness. Based on studies obtained by Rainnie et al. [52] it can be hypothesized that GABA interneurons are in close association with raphe serotonergic neurons innervating the amygdala, and remain under serotonergic control exerted through postsynaptic 5-HT2A receptors. Stimulation of these receptors, exclusively localized on GABAergic interneurons in the amygdala [53], leads to an increase in the frequency of inhibitory synaptic events in projection neurons, leading to inhibitory control over glutamatergic pyramidal neurons. Blockade of the 5-HT2A receptors leads to inhibition of the interneurons, and a lack of GABA inhibitory control over the glutamate network.

5.1.1 Clinical Data

Fenobam is a potent, selective, and noncompetitive mGlu5 receptor antagonist with an inverse agonist activity [13] that exhibited anxiolytic effect in the SIH model,

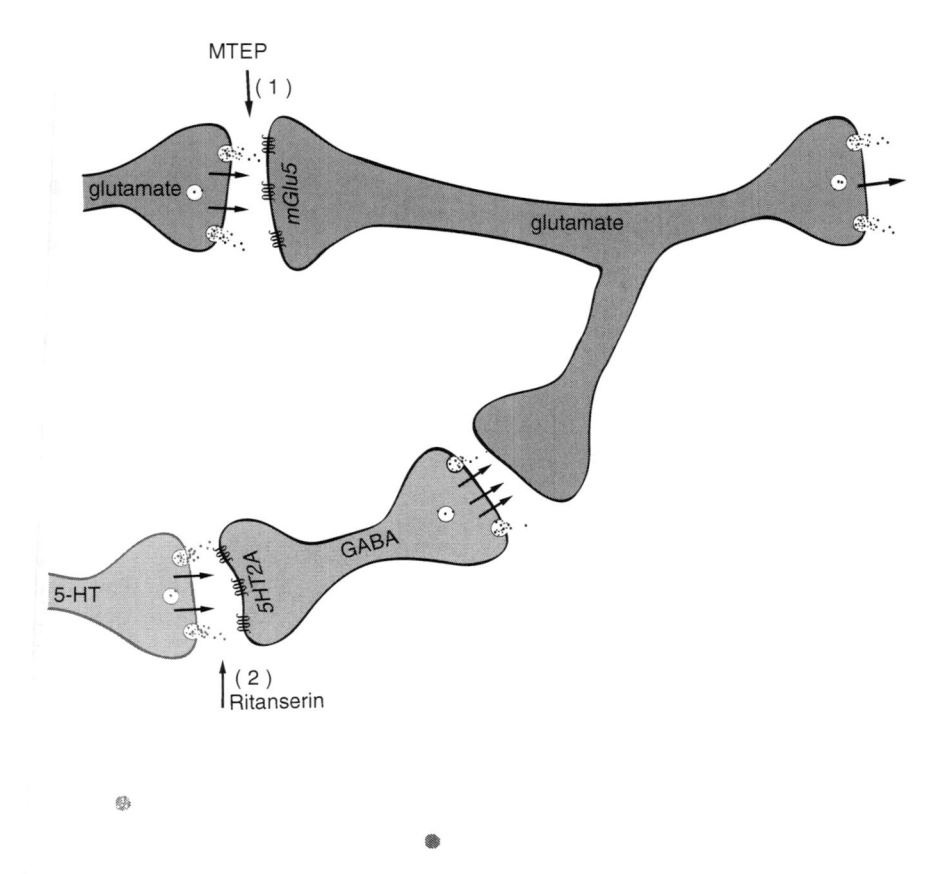

Fig. 2 The possible mechanism of anxiolytic-like effects of MTEP. The ligard evokes anxiolytic responses probably by decreasing glutamatergic transmission via inhibition of glutamate action on postsynaptic mGlu5 receptors (1). The 5-HT-GABA inhibitory action over glutamatergic network strongly contributes to the effect of MTEP, as the blockade of postsynaptic 5-HT2A receptors located on GABA-ergic interneurons prevents MTEP-induced anxiolysis (2). *One arrow*: transmission decreased, *two arrows*: normal transmission; *three arrows*: transmission increased

Vogel conflict test, GST, and conditioned emotional response test [13]. Importantly, fenobam is a clinically validated nonbenzodiazepine anxiolytic, introduced to clinical practice in 1980 (see [54]). The non-GABAergic activity of fenobam, together with its robust anxiolytic activity and reported efficacy in humans in a double blind placebo-controlled trial, supports the potential of developing mGlu5 receptor antagonists as novel anxiolytic agents, with an improved therapeutic window over benzodiazepines. Fenobam was free of adverse effects such as sedation, muscle relaxation, and interactions with alcohol [55]. The trials were, however, discontinued since it produced psychostimulant adverse effects in some patients [54]. The psychostimulant effects of fenobam might represent "off-target" activity of the compound, perhaps at the dopamine transporter.

5.2 Group II mGlu Receptors

Ligands of the group II mGlu receptors, mainly agonists and positive modulators, exert strong anxiolytic-like activity in standard animal models (for review see [21, 22, 26, 35, 36]). In contrast to the mGlu5 receptor modulators, ligands that activated group II mGlu receptors (such as DCG-IV, LY354740) were shown to impair consolidation of fear conditioning with no influence on preconditioning startle response [56, 57]. Parallel electrophysiological experiments revealed depotentiation or direct inhibition of the excitatory drive of LA synapses after activation of the group II mGlu receptors [56, 58].

The mechanism of action is based on their ability to inhibit glutamate release, although the precise localization of the receptors in the synaptic cleft is still not clear. Pre- and postsynaptic localization was reported, but the exact combination of mGlu2 and mGlu3 in a single synapse is not definitely known. Generally, mGlu2 receptors were shown as being expressed both on the dendritic shafts and along the axon terminals, where they modulate the release of the neurotransmitter at presynaptic terminals. Postsynaptic localization of mGlu2 receptors was described mainly for glutamatergic neurons, while the majority of presynaptic localization was observed on the GABAergic interneurons [59], although their expression as autoreceptors on excitatory terminals was also noticed [33]. Therefore, mGlu2/3 receptors can regulate the release of both excitatory and inhibitory amino acids and, depending on the type of the synapse they create, their modulators can influence the balance of the neurotransmitters in different ways. As mGlu2/3 receptors were found to be located both on the terminals of asymmetrical (glutamate–glutamate) or symmetrical (GABA–GABA) synapses, the putative mechanism of action of their activators could be as follows:

The inhibition of glutamate release in conventional glutamatergic synapses leads to a decreased excitation of the postsynaptic neuronal element, usually overactive in an anxious state. On the other hand, the inhibition of the release of the inhibitory neurotransmitter in the symmetrical synapses leads to disinhibition of the GABAergic neuron projection. The increased level of GABA released by the interneuron innervating pyramidal cells abolishes the overexcitation induced by anxiety, thus contributing to anxiolysis (Fig. 3).

The process of anxiolysis mediated by the mGlu2/3 receptors' ligands was reported to be independent of benzodiazepine receptors, similar to MTEP [57, 60]. The involvement of NPY signaling in the L-CCG-I effect was reported [61], whilst the influence of the other neurotransmitters system remains to be established.

Discussing the role of mGlu2/3 receptors in the mechanism of the anxiolytic action of their modulators, the role of astrocytes cannot be completely omitted, as the mGlu3 subtype was shown to be highly expressed on their surface [21]. mGlu3 receptors on astrocytic membranes wrapping the synapses may enhance conversion of glutamate into glutamine and thus contribute to the control of the glutamate level in the synaptic cleft [33]. It is difficult to presently speculate if this event has any contribution to the anxiolytic action of mGlu2/3 agonists.

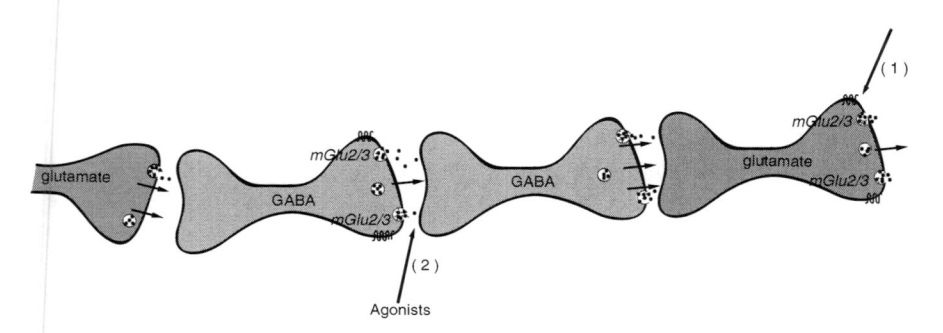

Fig. 3 The possible mechanism of anxiolytic-like effects of group II agonists/positive modulators. The ligands can evoke anxiolytic responses decreasing glutamatergic transmission by inhibition of glutamate release via stimulation of presynaptic group II mGlu autoreceptors (1) or by decreasing GABAergic transmission in symmetrical synapses. (2) As the result the increase in GABAergic transmission can occur due to inhibition of GABA release from the GABAergic interneuron via presynaptic group II mGlu heteroreceptors leading to disinhibition of the GABAergic projection neuron (2). *One arrow*: transmission decreased, *two arrows*: normal transmission; *three arrows*: transmission increased

5.2.1 Clinical Studies

Studies of patients with panic disorder revealed that LY354740 failed to produce any significant effects after 3 weeks of treatment different from the placebo [62]. Problems with the efficacy of LY354740 in humans resulted probably from a low bioavailability of the drug following oral administration because of poor gastrointestinal absorption [63]. The bioavailability of LY354740 was improved by the formation of an N dipeptide derivative of LY354740, named LY544344, which is actively transported from the gastrointestinal tract via the peptide transporter PepT1 64. After absorption, LY544344 is hydrolyzed to alanine and an active mGlu2/3 receptor agonist, LY354740. Data showed an 8–10-fold increase in the absorption and up to a 300-fold increase during in vivo activity (including anxiolytic-like effects in rats and/or mice) of LY544344, which can be compared with LY354740 [64, 65]. LY544344 was used in phase III clinical studies for the treatment of generalized anxiety disorder [12]. The placebo-controlled clinical trial showed significantly greater improvement from baseline in Hamilton Anxiety and Clinical Global Impression–
Improvement scores, as well as response and remission rates that compared with placebo-treated patients. LY544344 was well tolerated and there were no significant differences in the incidence of treatment-emergent adverse events among the three treatment groups. However, the trial was discontinued because of discouraging results showing seizures after chronic treatment with high doses of LY544344 in mice, rats, and dogs. However, such effects were supposed to result from the off-target activity of the compound and were observed only after repeated treatment with very high doses [66]. In conclusion, the findings of this study support the potential efficacy of the mGlu2/3 receptor agonist agents in the treatment of GAD.

Therefore, derivatives of LY354740 or other group II mGlu receptor agonists still seem to be promising potential anxiolytics. But the potential application of these compounds will depend on the results of further clinical studies.

Another selective agonist for metabotropic glutamate 2/3 (mGlu2/3) receptors is LY404039. An oral prodrug of LY404039 (LY2140023) was evaluated in schizophrenic patients showing statistically significant improvements in both positive and negative symptoms of schizophrenia compared to the placebo [67], although it is surprising that no data on the potential anxiolytic activity of that compound was reported.

Ultimately, positive allosteric modulators of mGlu2 receptors, inducing robust anxiolytic-like effects in preclinical studies, provide a new approach to the development of novel anxiolytics. However, more preclinical and thorough clinical studies are necessary to confirm such conclusions.

5.3   Group III mGlu Receptors

The third group of mGlu receptors constitutes the largest family; all of the subtypes together with their splice variants form a family of nearly 12 proteins. All the receptors are highly expressed throughout the CNS with the exception of the mGlu6 subtype, where the distribution was found to be restricted only to the retina (for review see [21, 22, 26, 35, 36]). Anxiolytic-like activity was described for selective mGlu4 and mGlu7 receptors positive allosteric modulators [68, 69] as well as for the nonselective agonist of mGlu4/8 receptors subtypes, ACPT-I [70]. AMN082 a positive allosteric modulator of mGlu7 receptors as well as an orthosteric agonist of the mGlu8 receptor, (S)-3,4-DCPG, were shown to be active in the FPS paradigm, retarding acquisition of CF [71, 72]. (S)-3,4-DCPG affected not only the acquisition but also the expression of conditioned fear in a manner similar to commonly used anxiolytic drugs [71], although no anxiolytic action was observed for the mGlu8 agonist in Vogel's test [31]. The behavioral results are in agreement with functional studies indicating the attenuation of LTP at thalamo–amygdalar synapses in the LA by AMN082 [71]. (S)-3,4-DCPG, in contrast, had a strong inhibitory effect on basal synaptic transmission and only a weak influence on LTP. The most likely mechanism of action of AMN082 is schematically shown in Fig. 4a. The privileged localization of these receptors is on the axon terminals of GABAergic neurons innervating interneurons. Thus, their stimulation leads to the inhibition of GABA release from presynaptic terminals and the disinhibition of postsynaptic inhibitory interneuron. An inhibition of inhibitory neurons leads to overinhibition of the pyramidal network (described also for group II ligands, see above) and thus establishes the function of the glutamate efferent at the normal level. The efficacy of these ligands, at least AMN082, was shown to be dependent on GABA signaling acting through $GABA_A$ receptors located on the pyramidal postsynaptic element (Fig. 4a).

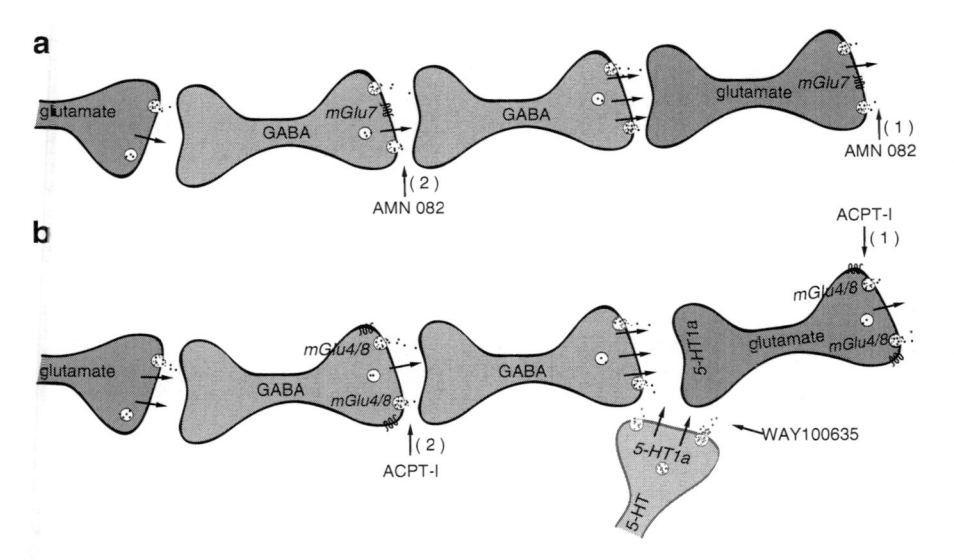

Fig. 4 (a) A possible mechanism for the anxiolytic-like effects of AMN082. AMN082 can evoke anxiolytic responses, decreasing glutamatergic transmission by inhibition of glutamate release via stimulation of presynaptic mGlu7 receptors (1) or by an increase in GABAergic transmission. The increase in GABAergic transmission can occur due to inhibition of GABA release from the GABAergic interneuron via presynaptic group III mGlu heteroreceptors leading to disinhibition of the GABAergic projection neuron (2), which controls glutamate activity. *One arrow*: transmission decreased, *two arrows*: normal transmission; *three arrows*: transmission increased. (**b**) The possible mechanism for the anxiolytic-like effects of ACPT-I. ACPT-I can evoke anxiolytic responses by decreasing glutamatergic transmission by inhibition of glutamate release via stimulation of presynaptic group III mGlu receptors (1) or by an increase in GABAergic transmission (2) in a manner similar to AMN082 described above. The parallel inhibition of glutamatergic neurons by serotonin is indispensable for anxiolytic action exerted by ACPT-I, because blockade of postsynaptic 5-HT1a receptors located on pyramidal neurons abolishes the effect of ACPT-I (2). *One arrow*: transmission decreased, *two arrows*: normal transmission; *three arrows*: transmission increased

The involvement of other neurotransmitter systems in the anxiolytic action of group III receptors activators was shown for ACPT-I, a nonselective agonist, which binds to mGlu4 and mGlu8 receptors with equal potency. The affinity of ACPT-I for mGlu7 subtype is relatively weak and can be ignored when considering the action of the ligand [26]. It was shown that the selective mGlu8 receptor agonist, (S)-3,4-DCPG, had no effect on unlearned fear in rats [31] and it, therefore, seems the agonist action on the mGlu4 receptor subtype is a key component of ACPT-I-mediated anxiolysis. A schematic diagram of the mechanism of action of ACPT-I is represented in Fig. 4b. Activation of mGlu4/8 autoreceptors reduces glutamate release from the presynaptic terminal, normalizing the function on glutamatergic synapses on one side. Second, activation of the heteroreceptors localized on interneurons of symmetrical synapses may lower the inhibitory action on GABAergic neurons, which themselves control the activity of the glutamatergic amygdalar output. This process was shown to be dependent on serotonergic system activity [70]. WAY100635, an antagonist of 5-HT1a receptors, blocked the effect of ACPT-I in the stress-induced hyperthermia test [70]. 5-HT1a receptors can be localized both pre and postsynaptically, and are dense in the amygdala, especially in the CeA nucleus [73]. 5-HT efflux in the central amygdala was shown to be under inhibitory control of 5-HT1a receptors in raphe nuclei [74]. The serotonin released from 5-HT terminals in the CeA acting through 5-HT1a receptors localized on the glutamatergic postsynaptic element may contribute to the inhibition of increased firing of these neurons. Systemic administration of WAY 100635 alone had no effect on the 5-HT1a release in the CeA, but it completely abolished the decrease in serotonin release induced by 5-HT1a agonists [75]. The blockade by WAY100653 of 5-HT1a postsynaptic receptors can block the inhibitory effect of 5-HT on pyramidal neurons, having no effect on 5-HT release per se.

6 Conclusions

Emerging evidence suggests that glutamate plays an important role in the pathophysiology of anxiety disorders. Preclinical data demonstrates that the agonists of group II and III mGlu receptors, as well as antagonists of group I mGlu receptors, can be future anxiolytic drugs. The data is corroborated by a limited amount of clinical data on group I and II mGlu receptor ligands. The anxiolytic action of mGlu receptor ligands involves complex neuronal networks consisting of at least glutamatergic, GABAergic, and serotonergic neurons in the amygdala and other brain structures.

References

1. Andlin-Sobocki P, Wittchen HU (2005) Cost of anxiety disorders in Europe. Eur J Neurol 12 (Suppl 1):39–44
2. DuPont RL, Rice DP, Miller LS, Shiraki SS, Rowland CR, Harwood HJ (1996) Economic costs of anxiety disorders. Anxiety 2:167–172
3. Greenberg PE, Sisitsky T, Kessler RC, Finkelstein SN, Berndt ER, Davidson JR, Ballenger JC, Fyer AJ (1999) The economic burden of anxiety disorders in the 1990s. J Clin Psychiatry 60:427–435
4. Javitt DC (2004) Glutamate as a therapeutic target in psychiatric disorders. Mol Psychiatry 9:979, 984–997
5. Cortese BM, Phan KL (2005) The role of glutamate in anxiety and related disorders. CNS Spectr 10:820–830
6. Bak LK, Schousboe A, Waagepetersen HS (2006) The glutamate/GABA-glutamine cycle: aspects of transport, neurotransmitter homeostasis and ammonia transfer. J Neurochem 98:641–653
7. Dinan T (2006) Therapeutic options: addressing the current dilemma. Eur Neuropsychopharmacol 16(Suppl 2):S119–S127

8. Ball SG, Kuhn A, Wall D, Shekhar A, Goddard AW (2005) Selective serotonin reuptake inhibitor treatment for generalized anxiety disorder: a double-blind, prospective comparison between paroxetine and sertraline. J Clin Psychiatry 66:94–99

9. Tokarski K, Bobula B, Wabno J, Hess G (2008) Repeated administration of imipramine attenuates glutamatergic transmission in rat frontal cortex. Neuroscience 153:789–795

10. Golembiowska K, Dziubina A (2000) Effect of acute and chronic administration of citalopram on glutamate and aspartate release in the rat prefrontal cortex. Pol J Pharmacol 52:441–448

11. Zarate CA Jr, Du J, Quiroz J, Gray NA, Denicoff KD, Singh J, Charney DS, Manji HK (2003) Regulation of cellular plasticity cascades in the pathophysiology and treatment of mood disorders: role of the glutamatergic system. Ann N Y Acad Sci 1003:273–291

12. Dunayevich E, Erickson J, Levine L, Landbloom R, Schoepp DD, Tollefson GD (2008) Efficacy and tolerability of an mGlu2/3 agonist in the treatment of generalized anxiety disorder. Neuropsychopharmacology 33:1603–1610

13. Porter RH, Jaeschke G, Spooren W, Ballard TM, Büttelmann B, Kolczewski S, Peters JU, Prinssen E, Wichmann J, Vieira E, Mühlemann A, Gatti S, Mutel V, Malherbe P (2005) Fenobam: a clinically validated nonbenzodiazepine anxiolytic is a potent, selective, and noncompetitive mGlu5 receptor antagonist with inverse agonist activity. J Pharmacol Exp Ther 315:711–721

14. Shin LM, Liberzon I (2010) The neurocircuitry of fear, stress, and anxiety disorders. Neuropsychopharmacology 35:169–191

15. Sah P, Westbrook RF (2008) Behavioural neuroscience: the circuit of fear. Nature 454:589–590

16. Harmer CJ, Mackay CE, ReidCB CPJ, Goodwin GM (2006) Antidepressant drug treatment modifies the neural processing of nonconscious threat cues. Biol Psychiatry 59:816–820

17. Paulus MP, Feinstein JS, Castillo G, Simmons AN, Stein MB (2005) Dose-dependent decrease of activation in bilateral amygdala and insula by lorazepam during emotion processing. Arch Gen Psychiatry 62:282–288

18. Davis M, Rainnie D, Cassell M (1994) Neurotransmission in the rat amygdala related to fear and anxiety. Trends Neurosci 17:208–214

19. Davis M, Whalen PJ (2001) The amygdala: vigilance and emotion. Mol Psychiatry 6:13–34

20. Amunts K, Kedo O, Kindler M, Pieperhoff P, Mohlberg H, Shah NJ, Habel U, Schneider F, Zilles K (2005) Cytoarchitectonic mapping of the human amygdala, hippocampal region and entorhinal cortex: intersubject variability and probability maps. Anat Embryol (Berl) 210:343–352

21. Wieronska JM, Pilc A (2009) Metabotropic glutamate receptors in the tripartite synapse as a target for new psychotropic drugs. Neurochem Int 55:85–97

22. Palucha A, Pilc A (2007) Metabotropic glutamate receptor ligands as possible anxiolytic and antidepressant drugs. Pharmacol Ther 115:116–147

23. Pin JP, De Colle C, Bessis AS, Acher F (1999) New perspectives for the development of selective metabotropic glutamate receptor ligands. Eur J Pharmacol 375:277–294

24. Schoepp DD, Jane DE, Monn JA (1999) Pharmacological agents acting at subtypes of metabotropic glutamate receptors. Neuropharmacology 38:1431–1476

25. Lujan R, Roberts JD, Shigemoto R, Ohishi H, Somogyi P (1007) Differential plasma membrane distribution of metabotropic glutamate receptors mGluR1 alpha, mGluR2 and mGluR5, relative to neurotransmitter release sites. J Chem Neuroanat 13:219–241

26. Lavreysen H, Dautzenberg FM (2008) Therapeutic potential of group III metabotropic glutamate receptors. Curr Med Chem 15:671–684

27. Somogyi P, Dalezios Y, Luján R, Roberts JD, Watanabe M, Shigemoto R (2003) High level of mGluR7 in the presynaptic active zones of select populations of GABAergic terminals innervating interneurons in the rat hippocampus. Eur J Neurosci 17:2503–2520

28. Kogo N, Dalezios Y, Capogna M, Ferraguti F, Shigemoto R, Somogyi P (2004) Depression of GABAergic input to identified hippocampal neurons by group III metabotropic glutamate receptors in the rat. Eur J Neurosci 19:2727–2740

29. Wieronska JM, Smiałowska M, Brański P, Gasparini F, Kłodzińska A, Szewczyk B, Pałucha A, Chojnacka-Wójcik E, Pilc A (2004) In the amygdala anxiolytic action of mGlu5 receptors antagonist MPEP involves neuropeptide Y but not GABAA signaling. Neuropsychopharmacology 29:514–521

30. ShigemotoR KA, Wada E, Nomura S, Ohishi H, Takada M, Flor PJ, Neki A, Abe T, Nakanishi S, Mizuno N (1997) Differential presynaptic localization of metabotropic glutamate receptor subtypes in the rat hippocampus. J Neurosci 17:7503–7522

31. Stachowicz K, Klak K, Pilc A, Chojnacka-Wojcik E (2005) Lack of the antianxiety-like effect of (S)-3, 4-DCPG, an mGlu8 receptor agonist, after central administration in rats. Pharmacol Rep 57:856–860

32. Smialowska M, Szewczyk B, Brański P, Wierońska JM, Pałucha A, Bajkowska M, Pilc A (2002) Effect of chronic imipramine or electroconvulsive shock on the expression of mGluR1a and mGluR5a immunoreactivity in rat brain hippocampus. Neuropharmacology 42:1016–1023

33. Petralia RS, Wang YX, Niedzielski AS, Wenthold RJ (1996) The metabotropic glutamate receptors, mGluR2 and mGluR3, show unique postsynaptic, presynaptic and glial localizations. Neuroscience 71:949–976

34. Kinoshita A, Shigemoto R, Ohishi H, van der Putten H, Mizuno N (1998) Immunohistochemical localization of metabotropic glutamate receptors, mGluR7a and mGluR7b, in the central nervous system of the adult rat and mouse: a light and electron microscopic study. J Comp Neurol 393:332–352

35. Swanson CJ, Bures M, Johnson MP, Linden AM, Monn JA, Schoepp DD (2005) Metabotropic glutamate receptors as novel targets for anxiety and stress disorders. Nat Rev Drug Discov 4:131–144

36. Chojnacka-Wojcik E, Klodzinska A, Pilc A (2001) Glutamate receptor ligands as anxiolytics. Curr Opin Investig Drugs 2:1112–1119

37. Shipe WD, Wolkenberg SE, Williams DL Jr, Lindsley CW (2005) Recent advances in positive allosteric modulators of metabotropic glutamate receptors. Curr Opin Drug Discov Devel 8:449–457

38. Slassi A, Isaac M, Edwards L, Minidis A, Wensbo D, Mattsson J, Nilsson K, Raboisson P, McLeod D, Stormann TM, Hammerland LG, Johnson E (2005) Recent advances in noncompetitive mGlu5 receptor antagonists and their potential therapeutic applications. Curr Top Med Chem 5:897–911

39. Marino MJ, Conn PJ (2006) Glutamate-based therapeutic approaches: allosteric modulators of metabotropic glutamate receptors. Curr Opin Pharmacol 6:98–102

40. Bach P (2007) Metabotropic glutamate receptor 5 modulators and their potential therapeutic applications. Expert Opin Ther Pat 17:371–384

41. Jaeschke G (2008) mGlu5 receptor antagonists and their therapeutic potential. Expert Opin Ther Pat 18:123–142

42. Millan MJ (2003) The neurobiology and control of anxious states. Prog Neurobiol 70:83–244

43. Fendt M, Fanselow MS (1999) The neuroanatomical and neurochemical basis of conditioned fear. Neurosci Biobehav Rev 23:743–760

44. Clugnet MC, LeDoux JE (1990) Synaptic plasticity in fear conditioning circuits: induction of LTP in the lateral nucleus of the amygdala by stimulation of the medial geniculate body. J Neurosci 10:2818–2824

45. Maren S, Fanselow MS (1995) Synaptic plasticity in the basolateral amygdala induced by hippocampal formation stimulation in vivo. J Neurosci 15:7548–7564

46. Li XF, Phillips R, LeDoux JE (1995) NMDA and non-NMDA receptors contribute to synaptic transmission between the medial geniculate body and the lateral nucleus of the amygdala. Exp Brain Res 105:87–100

47. Pilc A, Kłodzińska A, Brański P, Nowak G, Pałucha A, Szewczyk B, Tatarczyńska E, Chojnacka-Wójcik E, Wierońska JM (2002) Multiple MPEP administrations evoke anxiolytic- and antidepressant-like effects in rats. Neuropharmacology 43:181–187

48. Jaeschke G, Porter R, Büttelmann B, Ceccarelli SM, Guba W, Kuhn B, Kolczewski S, Huwyler J, Mutel V, Peters JU, Ballard T, Prinssen E, Vieira E, Wichmann J, Spooren W (2007) Synthesis and biological evaluation of fenobam analogs as mGlu5 receptor antagonists. Bioorg Med Chem Lett 17:1307–1311

49. Fendt M, Schmid S (2002) Metabotropic glutamate receptors are involved in amygdaloid plasticity. Eur J Neurosci 15:1535–1541

50. Klodzinska A, Tatarczyńska E, Stachowicz K, Chojnacka-Wójcik E (2004) Anxiolytic-like effects of MTEP, a potent and selective mGlu5 receptor agonist does not involve GABA(A) signaling. Neuropharmacology 47:342–350

51. Stachowicz K, Gołembiowska K, Sowa M, Nowak G, Chojnacka-Wójcik E, Pilc A (2007) Anxiolytic-like action of MTEP expressed in the conflict drinking Vogel test in rats is serotonin dependent. Neuropharmacology 53:741–748

52. Rainnie DG (1999) Serotonergic modulation of neurotransmission in the rat basolateral amygdala. J Neurophysiol 82:69–85

53. Jiang X (2009) Stress impairs 5-HT2A receptor-mediated serotonergic facilitation of GABA release in juvenile rat basolateral amygdala. Neuropsychopharmacology 34:410–423

54. Pecknold JC, McClure DJ, Appeltauer L, Wrzesinski L, Allan T (1982) Treatment of anxiety using fenobam (a nonbenzodiazepine) in a double-blind standard (diazepam) placebo-controlled study. J Clin Psychopharmacol 2:129–133

55. Goldberg ME, Salama AI, Patel JB, Malick JB (1983) Novel non-benzodiazepine anxiolytics. Neuropharmacology 22:1499–1504

56. Lin CH, Lee CC, Huang YC, Wang SJ, Gean PW (2005) Activation of group II metabotropic glutamate receptors induces depotentiation in amygdala slices and reduces fear-potentiated startle in rats. Learn Mem 12:130–137

57. Tizzano JP, Griffey KI, Schoepp DD (2002) The anxiolytic action of mGlu2/3 receptor agonist, LY354740, in the fear-potentiated startle model in rats is mechanistically distinct from diazepam. Pharmacol Biochem Behav 73:367–374

58. Muly EC, Mania I, Guo JD, Rainnie DG (2007) Group II metabotropic glutamate receptors in anxiety circuitry: correspondence of physiological response and subcellular distribution. J Comp Neurol 505:682–700

59. Ohishi H, Nomura S, Ding YQ, Shigemoto R, Wada E, Kinoshita A, Li JL, Neki A, Nakanishi S, Mizuno N (1994) Immunohistochemical localization of metabotropic glutamate receptors, mGluR2 and mGluR3, in rat cerebellar cortex. Neuron 13:55–66

60. Linden AM, Shannon H, Baez M, Yu JL, Koester A, Schoepp DD (2005) Anxiolytic-like activity of the mGLU2/3 receptor agonist LY354740 in the elevated plus maze test is disrupted in metabotropic glutamate receptor 2 and 3 knock-out mice. Psychopharmacology (Berl) 179:284–291

61. Smialowska M, Wieronska JM, Domin H, Zieba B (2007) The effect of intrahippocampal injection of group II and III metobotropic glutamate receptor agonists on anxiety; the role of neuropeptide Y. Neuropsychopharmacology 32:1242–1250

62. Bergink V, Westenberg HG (2005) Metabotropic glutamate II receptor agonists in panic disorder: a double blind clinical trial with LY354740. Int Clin Psychopharmacol 20:291–293

63. Johnson JT, Mattiuz EL, Chay SH, Herman JL, Wheeler WJ, Kassahun K, Swanson SP, Phillips DL (2002) The disposition, metabolism, and pharmacokinetics of a selective metabotropic glutamate receptor agonist in rats and dogs. Drug Metab Dispos 30:27–33

64. Bueno AB (2005) Dipeptides as effective prodrugs of the unnatural amino acid (+)-2-aminobicyclo[3.1.0]hexane-2, 6-dicarboxylic acid (LY354740), a selective group II metabotropic glutamate receptor agonist. J Med Chem 48:5305–5320

65. Rorick-Kehn LM, Perkins EJ, Knitowski KM, Hart JC, Johnson BG, Schoepp DD, McKinzie DL (2006) Improved bioavailability of the mGlu2/3 receptor agonist LY354740 using a prodrug strategy: in vivo pharmacology of LY544344. J Pharmacol Exp Ther 316:905–913

66. Danysz W (2005) LY-544344. Eli Lilly. Drugs 8:755–762

67. Patil ST, Zhang L, Martenyi F, Lowe SL, Jackson KA, Andreev BV, Avedisova AS, Bardenstein LM, Gurovich IY, Morozova MA, Mosolov SN, Neznanov NG, Reznik AM, Smulevich AB, Tochilov VA, Johnson BG, Monn JA, Schoepp DD (2007) Activation of mGlu2/3 receptors as a new approach to treat schizophrenia: a randomized Phase 2 clinical trial. Nat Med 13:1102–1107

68. Stachowicz K, Klak K, Klodzinska A, Chojnacka-Wojcik E, Pilc A (2004) Anxiolytic-like effects of PHCCC, an allosteric modulator of mGlu4 receptors, in rats. Eur J Pharmacol 498:153–156

69. Stachowicz K, Brañski P, Kłak K, van der Putten H, Cryan JF, Flor PJ, Andrzej P (2008) Selective activation of metabotropic G-protein-coupled glutamate 7 receptor elicits anxiolytic-like effects in mice by modulating GABAergic neurotransmission. Behav Pharmacol 19:597–603

70. Stachowicz K, Kłodzińska A, Palucha-Poniewiera A, Schann S, Neuville P, Pilc A (2009) The group III mGlu receptor agonist ACPT-I exerts anxiolytic-like but not antidepressant-like effects, mediated by the serotonergic and GABA-ergic systems. Neuropharmacology 57:227–234

71. Fendt M, Schmid S, Thakker DR, Jacobson LH, Yamamoto R, Mitsukawa K, Maier R, Natt F, Hüsken D, Kelly PH, McAllister KH, Hoyer D, van der Putten H, Cryan JF, Flor PJ (2008) mGluR7 facilitates extinction of aversive memories and controls amygdala plasticity. Mol Psychiatry 13:970–979

72. Schmid S, Fendt M (2006) Effects of the mGluR8 agonist (S)-3, 4-DCPG in the lateral amygdala on acquisition/expression of fear-potentiated startle, synaptic transmission, and plasticity. Neuropharmacology 50:154–164

73. Ohuoha DC, Hyde TM, Kleinman JE (1993) The role of serotonin in schizophrenia: an overview of the nomenclature, distribution and alterations of serotonin receptors in the central nervous system. Psychopharmacology (Berl) 112:S5–S15

74. Bosker FJ, Klompmakers A, Westenberg HG (1997) Postsynaptic 5-HT1A receptors mediate 5-hydroxytryptamine release in the amygdala through a feedback to the caudal linear raphe. Eur J Pharmacol 333:147–157

75. Bosker F, Vrinten D, Klompmakers A, Westenberg H (1997) The effects of a 5-HT1A receptor agonist and antagonist on the 5-hydroxytryptamine release in the central nucleus of the amygdala: a microdialysis study with flesinoxan and WAY 100635. Naunyn Schmiedebergs Arch Pharmacol 355:347–353

Index